Conflict Prevention in Project Management

Wolfgang Spiess · Finn Felding

Conflict Prevention in Project Management

Strategies, Methods, Checklists and Case Studies

Dr. Wolfgang Spiess, consultant
Pappelstr. 5
46519 Alpen
Germany
wolf.spiess@t-online.de

Finn Felding, project adviser
Hvalsoevej 7, Osager
4320 Lejre
Denmark
ff@finnfelding.dk

ISBN 978-3-540-77436-5 e-ISBN 978-3-540-77437-2

DOI 10.1007/978-3-540-77437-2

Library of Congress Control Number: 2008929584

Cover design: Frido Steinen-Broo, eStudio Calamar, Spain
Production: le-tex publishing services oHG, Leipzig

Printed on acid-free paper

9 8 7 6 5 4 3 2 1

springer.com

Preface

This handbook has three primary objectives : (1) to give the project managers guidance to avoid conflicts in project execution and to understand the procedures in case of legal proceedings, (2) to give lawyers the understanding of the technical problems in project management, and (3) to give students an introduction into the technical and legal aspects of managing big international projects. The case studies and questions at the end of each chapter are especially directed to the student and the young project managers, who try to enter the ever more complicated world of managing international projects.

This book does not try to give legal advice, but it tries to help engineers and project managers how to thoroughly plan their project in order to avoid conflicts during execution. In this way it also helps lawyers to better understand their clients, when they have to defend them in conflicts regarding big international projects. The authors' many years of experience in managing international projects on one side and in assisting as experts and monitors of litigation on the other side have led them to write this book and thus to help other project managers avoid the mistakes that they themselves and other project managers have made in the past.

The book is organized into nine chapters and a glossary, the latter trying to give a short definition of the terms used in this book for easier understanding. Chapter 4 of this book is written by guest writer Rikke Rye Hahn, psychologist from Svendborg, Denmark (rikke_hahn@yahoo.dk). The authors thank Mrs. Hahn for her excellent contribution to this book.

The authors are pleased to thank the companies "Cimbria Unigrain A/S" (www.cimbria.com), "FLSmidth A/S" (www.flsmidth.com) and "Kosan Crisplant A/S" (www.kosancrisplant.com) for providing photos, which illustrate international projects of the type discussed in this book, although these "project photos" are in no way related to the case studies discussed in the book.

The authors especially thank Jenny and Dr. Brian Smith, who kindly did the proof reading of this book.

The pleasant cooperation with Springer-Verlag GmbH and especially Dr.-Ing. Boris Gebhardt, the Engineering Editor of Springer, has helped the authors to get through the not always easy task of preparing the book in a printable version: The authors thank Springer-Verlag for accepting their ideas in the planning of this book.

At a personal level the authors dedicate this book to their wives Lillian Felding and Nicole Delpiano Spiess and thank them for their patience and support, which made this book become reality.

Contents

1 Introduction 1

2 Parties, Roles and Interests in International Projects 5
- 2.1 The Parties 5
- 2.2 Distribution of Responsibilities Between Contracting Parties 7
- 2.3 Analysis of Contracting Parties and Their Interests 9
- 2.4 Conflict Contingency Plan 9
- 2.5 Laws, Regulations, Decrees and Standards Governing the Contract . 9
- 2.6 The Authorities of the Project Country 11
- 2.7 Other Players 12
- 2.8 Conclusion 12
- References 13

3 Engineering, Supply and Construction Contracts 15
- 3.1 Contract Types and Their Elements 16
- 3.2 Material Obligations and the Drafting Process 17
- 3.3 Regulatory Obligations 22
- 3.4 Financial Obligations 23
  - 3.4.1 Price and Payment Terms 23
  - 3.4.2 Security System 25
  - 3.4.3 Relation Between the Engineering, Supply and Construction Contract and the Project Finance Agreement 25
- 3.5 Taxes 26
- 3.6 Project Organisation Provisions 27
- 3.7 Dispute Resolution Provisions 27
- 3.8 Contract Drafting in Europe, Asia and America 28
  - 3.8.1 Introduction 28
  - 3.8.2 Survey on Lawyers' Opinion on Conflicts in International Projects 30
  - 3.8.3 Asian Contract Drafting as Observed by Europeans 34
  - 3.8.4 American Contract Drafting as Observed by Europeans 36
  - 3.8.5 European Contract Drafting as Observed by Europeans 36
  - 3.8.6 International Standards and Good Practises 37
- 3.9 Conclusion 38

3.10 Questions on Chapter 3 ... 39
References ... 40

4 **Preventing Conflicts by Application of Psychology** ... 41
4.1 Case Study: The Fertilizer Plant ... 41
4.2 Understanding the Dynamics of Conflicts ... 43
4.2.1 Level One (Win Win) ... 43
4.2.2 Level Two (Win Lost) ... 44
4.2.3 Level Three (Lost Lost) ... 45
4.2.4 Evaluation of the Friedrich Glasl Model with Respect to Conflict Escalation ... 45
4.3 Dimensions of Conflicts on Project Level ... 46
4.4 Understanding One's Own and the Other Party's Reactions to Conflicts ... 49
4.4.1 The Persons Involved ... 50
4.4.2 Different Types of Personalities ... 50
4.5 Fostering Constructive Responses to Conflicts ... 54
4.5.1 Typical Ways of Acting and Communicating when Facing Conflicts ... 54
4.5.2 Destructive and Constructive Conflict Communication ... 55
4.6 Understanding Organizational Differences ... 57
4.7 Suggestions About How to Work out a Psychological Contract ... 58
4.7.1 The "ADR-Clause" ... 58
4.7.2 Less Formal – the Internal Psychological Check List ... 59
4.8 Conclusion to Chapter 4 ... 59
4.9 Questions on Chapter 4 ... 60
References ... 60

5 **Negotiations Leading to Conflict Resolution** ... 63
5.1 Introduction ... 63
5.2 Conflict Causes and Sources ... 64
5.3 Why Commercial Negotiation is the Preferred Method of Conflict Prevention ... 68
5.4 How Can the Rate of Success in Commercial Negotiations Be Improved? ... 71
5.5 The Contract Parties and Their Situation ... 73
5.6 Preparation of Negotiations ... 73
5.6.1 Basics in Preparation of Negotiations ... 73
5.6.2 Negotiating Team ... 75
5.7 Suggestions on How to Start Settlement Negotiations ... 77
5.8 Negotiations Leading to Settlement ... 78
5.8.1 Basics of Negotiations ... 78
5.8.2 Attitudes and Behaviour of Negotiators ... 79
5.9 Making the Agreement for a Settlement ... 80
5.9.1 Basics of a Commercial Settlement ... 80

5.9.2 Final Settlement Discussions 80
5.9.3 Drafting the Settlement Agreement 81
5.9.4 Concluding and Signing the Settlement Agreement......... 82
5.10 Handling Break-down of Negotiations........................... 82
5.11 Negotiations Parallel with Litigation............................ 83
5.12 Executing the Settlement Agreement 83
5.13 Negotiation of Delays and Extension of Time 84
5.14 Regional Differences 87
5.15 Case Studies.. 88
5.15.1 The Raw Material Plant 89
5.15.2 The Mineral Processing Plant............................ 90
5.15.3 The Food Processing Plant 92
5.15.4 The New Technology Plant 93
5.15.5 The Semiconductor Project.............................. 94
5.16 Conclusion to Chapter 5.. 96
5.17 Questions on Chapter 5 97
References.. 98

6 Litigation, Arbitration and Mediation Contributing to Conflict Settlement 99
6.1 Project Manager and Monitor of Litigation 99
6.1.1 Comparison of the Tasks of Project Manager and Monitor of Litigation.. 99
6.1.2 The Monitor of Litigation 101
6.2 Considerations Concerning the Strategy for Litigation 102
6.3 Pre-Arbitral and Soft Resolution Methods 106
6.3.1 Mediation and Referees Stipulated in the Contract 106
6.3.2 The Corporate Pledge Model 108
6.4 Arbitration and Litigation 109
6.4.1 General Considerations 109
6.4.2 Arbitration.. 110
6.4.3 Litigation 113
6.4.4 The Choice of Arbitrators 113
6.4.5 The Cost of Litigation................................... 114
6.4.6 The Preparation of Evidence............................. 117
6.5 Case Studies and Questions....................................... 119
6.5.1 The Soft Drink Factory.................................. 119
6.5.2 The Alcohol Plant 120
6.6 Conclusion to Chapter 6 122
6.7 Questions on Chapter 6 122
References.. 123

7 Expertise Contributing to Conflict Solutions 125
7.1 The Appointment of the Expert by the Court 126
7.2 The Appointment of the Expert by One or Both Parties 129
7.3 Execution of the Expertise 130
7.4 Can an Expert Be Rejected? 132
7.5 Cost of an Expertise 133
7.6 Case Studies 135
7.6.1 The Tire Mounting Factory 135
7.6.2 The Flour Mill 137
7.7 Conclusions to Chapter 7 138
7.8 Questions on Chapter 7 139
References 139

8 Project Management Tools to Help Avoid Conflicts 141
8.1 The Detailed Project Plan 142
8.1.1 Project Description and Objectives 142
8.1.2 Specifications or Task Statement and Functional Guarantees 143
8.1.3 Work Breakdown Structure 144
8.1.4 Time Schedule (PERT) and Delay Penalties 146
8.1.5 Organisation, Staffing and Responsibility Matrix 153
8.1.6 Procurement and Subcontracting 153
8.1.7 Budget 154
8.1.8 Governing Laws and Standards 154
8.1.9 Project Reporting 155
8.1.10 Project Acceptance Procedure 156
8.1.11 Risk Analysis 157
8.2 The Personnel for the Project 160
8.2.1 The Recruitment of Project Managers 160
8.2.2 Team Building and Team Relations 163
8.3 Case Studies and Questions 164
8.3.1 The Spring Roll Machine 164
8.3.2 The EPS plant 165
8.4 Conclusions to Chapter 8 166
8.5 Questions on Chapter 8 167
References 168

9 Conclusions and Recommendations 169

Glossary 173

Index 183

1 Introduction

This handbook is directed to the Senior Management of a company, to Sales Representatives and Project Managers and their staff who are in a constant struggle to bring projects to a successful conclusion. It is also directed to lawyers who assist companies in their international projects for contract drafting and conflict resolution. This book is not a legal but a technical and managerial textbook, and it does not intend to give legal advice, but it intends to give the lawyers assistance in their work from a managerial point of view.

Projects in conflict are not only causing losses, problems and bad reputation for the parties involved i.e. Clients, Consultants, Suppliers and Contractors but they can also have negative impacts on the economy of the project country.

In all fairness, conflicts between the parties are not the only reason for unsuccessful projects. The other reasons seem to be a lack of experience, not enough competence and insufficient preparation of the acting people. Such reasons are also bound to lead to unsuccessful projects and can also be the causes of conflicts during project execution.

It is our intention to describe and analyze how disagreements in big projects start between client and contractor, and may develop into major conflicts causing delays, overruns and inadequate quality of project execution. Therefore we will try to give advice on how to avoid conflicts and how to resolve them, in order to mitigate the losses in a project.

The economic consequences of major projects in conflict are sometimes disastrous, a company might go into bankruptcy or might require major fresh capital invested. In the case of smaller countries or countries in development the failure of big projects might even have a negative impact on their economic and industrial development and their growth.

Chapter 2 of this book describes the parties, which have an influence on the success of a project. These parties are the client, the contractor, the subcontractors and the surrounding entities such as the Government of the country of execution, other governing bodies, the unions, etc. The understanding of the roles and interests of these parties involved is necessary, in order to early recognize potential conflicts and to be able to handle them.

In Chap. 3 we discuss the different types of contracts for major international projects and the provisions, that should be agreed upon to lay the basis for a successful contract execution. Pitfalls and reasons for conflicts are discussed.

We shall not give legal advice, this is left to the lawyers, but we will show Project Managers, how to approach a new project, what to look for, when drawing up a contract together with the lawyers and which tools to use when starting and executing the project.

Chapter 4, written by psychologist Rikke Rye Hahn of Svendborg, Denmark, focuses on the psychological aspects, which play a role, when projects and the acting people are in conflict. Ways how to analyze conflicts are described and methods given, which help to handle and resolve conflicts. The "conflict competent leader" is described and "the psychological contract" is recommended as method to be utilized.

In Chap. 5 we will look at ways and means of how to negotiate a commercial settlement between the parties, in order to avoid legal proceedings. Sometimes little changes in the approach, in the persons acting, and in the timing of such negotiations may help to solve problems without going to court immediately. We present a number of methods to facilitate the commercial settlement process and a number of case studies to illustrate our main recommendations.

If the conflict cannot be solved "peacefully" via negotiations between the parties, and if it is taken to court or to arbitration, then our Chap. 6 will give advice of how Project Management should prepare and handle their case. The various existing procedures, as there are litigation in front of a state court, arbitration under the control of one of the existing arbitration institutions, mediation under assistance of a mediator, and the US Corporate Pledge Model are discussed. Case studies are presented, which elaborate on the acting peoples' approach with respect to the methods discussed in this chapter.

Since much litigation and arbitration has its cause in technical matters, where the right expertise is needed, we felt it useful to add a Chap. 7 on "Expertise" which is performed by an outside expert nominated by the Court or the parties. Many court cases have eventually been decided in their essence by the expert, who gave the basis for the judges' decision. Suggestions are made, how the parties should act with respect to an expertise, how experts should be chosen and how an expert could eventually be rejected.

Finally Chap. 8 describes and explains project management tools assisting the Project Manager to successfully prepare and execute his project without running into genuine conflicts. The well known methods, such as project plan, work breakdown structure, time schedule (PERT), organization and staffing, procurement and subcontracting, budget and governing laws and standards, project reporting, acceptance procedure and finally risk analysis are discussed not as a basic lecture in project management but how they should be used in order to avoid conflicts. Case studies and questions to the reader at the end of the chapter round up the professional use of project management tools.

In Chap. 9 we summarize the main conclusions and recommendations for better conflict prevention and handling in the management of projects.

Finally this handbook contains a glossary of the most important terms, which were discussed in this book, together with their explanations.

We hope that this handbook can contribute to a more productive and progressive project implementation with less disagreements or conflicts between the parties. The proposals made will cost extra resources in the early stages of a project, but will save a lot more resources by the end. Most important, though, the implementation of the proposed methods will result in a safer and more reliable project execution with less delays and rare budget overruns.

Another important benefit from a reduction in project conflicts is, that better qualified project people will be attracted to work for international projects. This will further improve the execution of the projects.

If project conflict handling and infighting are the major preoccupations of Project Managers, this key post will be held by introverted "bureaucrats" instead of extrovert globally oriented business managers. Project business needs smooth implementation with as little conflicts as possible.

We wish the reader pleasure in reading this handbook and a successful implementation of the methods and tools presented herein in his daily work.

The reader's suggestions and observations are very welcome to the following e-mail addresses:

Dr.-Ing. Wolfgang Spiess
Consultant and sworn expert

wolf.spiess@t-online.de

Finn Felding,
Project Adviser

ff@finnfelding.dk

2 Parties, Roles and Interests in International Projects

Abstract. Chapter 2 describes the parties, their interests, their responsibilities and their relations in international projects. It focuses on the parties' responsibilities and the related scope of work and services that totally should match the needs of the project implementation until successful commercial operation has been achieved.

As all international projects are implemented by several parties, the distribution of responsibilities and the plan on how to implement them in a coordinated manner is essential for preventing conflicts. In this respect knowledge about and understanding of the parties, their interests and their representatives play an important and underestimated role. This also includes the relation between the project and the local authorities in the project country.

Therefore a conflict contingency plan is recommended. It highlights the potential conflict areas of the project and prepares the parties for handling disagreements that can evolve into a conflict, thereby preventing the conflict. This Chap. 2 is also an introduction to Chap. 3 about the drafting of the project contract.

Key words: Project Parties; Parties' Interests; Parties' Responsibilities; Scope of Work; Owner (Client, Employer); Quality, Health, Safety & Environment (QHSE); Project Country; Local Authorities; Laws, Regulation, Decrees and Standards; Main Contractor; Main Process Supplier; Conflict Contingency Plan; Contracting Structure; Project Plan

2.1 The Parties

Turner (1993) has identified and classified the parties involved in a project in the following general manner:

a. "the parent organization (the owner of the facility),"
b. "the users, who will operate the facility"
c. "the supporters, who will supply the resources to undertake the work"
d. "the stakeholders, who are affected by the project"

Turner (1993) uses the Channel Tunnel project as an example of the parties involved in a international infrastructure project as shown in Table 2.1 below.

During the development, decision making and approval stages of major infrastructure or industrial projects conflicts of interests will often appear regarding environmental issues. These are outside the scope of this book because some of the parties are not contract parties but regulatory and political players.

Table 2.1. The parties involved in the Channel Tunnel project according to Turner (1993)

Role	Position	Group
Owner		Eurotunnel ; its shareholders
User	Operator	Eurotunnel
Manager		Trans Manche Link
Supporters	Financiers, Subcontractors, Project auditors, Suppliers	Banks world-wide, Partners in TML consortium, W.S. Atkins, Brittish Rail and SNCF
Stakeholders	Buyers, Competitors, Communities	Travelling public, hauliers, Cross-Channel ferries, London, Kent, Pas de Calais

International projects are normally implemented by a number of contracts thereby involving many contract parties bound together by a contracting structure. There are two basic types of contracting structure: A. "The main contract type" and B. "The parallel contracting by trade type" (i.e. civil, building, mechanical & electrical, refrigeration, heat, ventilation & air conditioning etc.) as illustrated by Figs. 2.1 and 2.2.

Everyone involved in a project in any major role needs to have an understanding of the parties, their objectives, their interests, their organization, their representatives and their expected behavior in order to provide the best performance during project implementation. This involves a great number of negotiations and compromises – especially for the representatives of the Owner, the main contractor and the main process supplier.

It is very important that all parties have a clear picture of this distribution of work and responsibilities in their mind and hopefully all are seeing the same picture and respecting it.

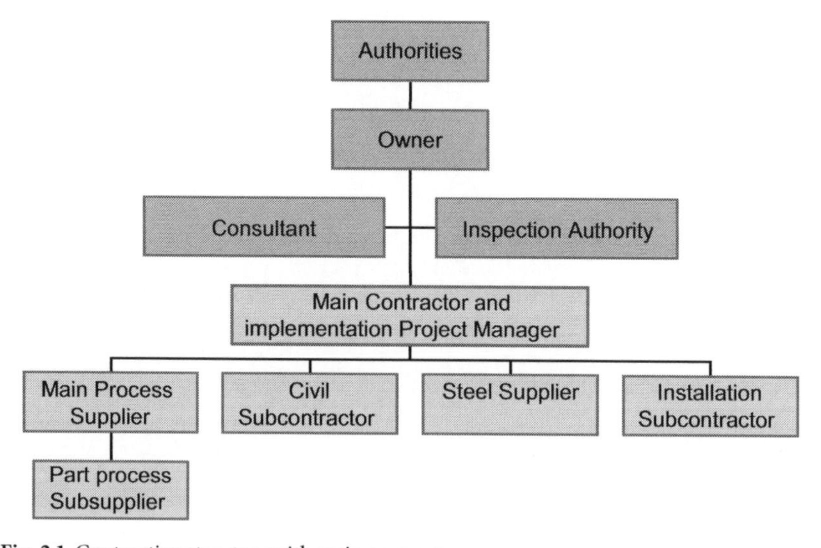

Fig. 2.1. Contracting structure with main contractor

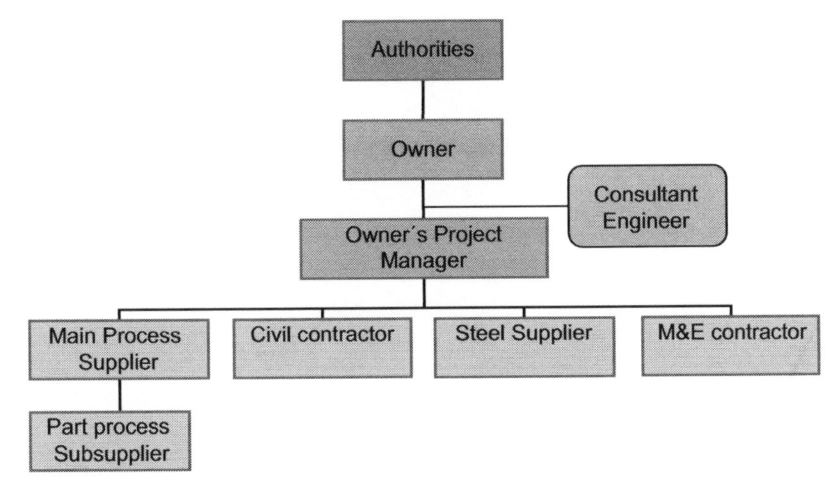

Fig. 2.2. Contracting Structure with parallel contractors

2.2 Distribution of Responsibilities Between Contracting Parties

The total scope of work and services necessary for the implementation of the project has to be divided between the suppliers, contractors and the Owner in a way that the total sum of the work packages equals the work necessary for the implementation of the project. Such a control of the distribution of the scope of work is recommended at the start of each project stage and should be performed jointly by the owner and the main contractor. Tables 2.2 and 2.3 below illustrate the typical distribution of responsibilities in the project development stage and in the implementation stages of engineering and construction.

In major international projects the relation between the Owner/the User on one side and the Main Process Supplier/Main Contractor on the other is of paramount importance.

Table 2.2. Distribution of scope of work in principle between Owner/User and Main Contractor/Main Process Supplier during the Project Development Stage

Task	Owner/User	Main Contractor/Main Process Supplier
Basic Concept	Main task holder	Advisory role
Process design	Defining input and wanted output	Main responsibility
Location	Main task holder	Advisory role/Expert role
General lay out	Main task holder	Expert role
Government Permits	Main task holder	Advisory role
Basic Work Breakdown Structure	Approval	Expert role
Contracting Structure	Approval	Expert role
Control of division of scope of work	Approval	Expert role

Table 2.3. Distribution of scope of work in principle between Owner/User and Main Contractor/ Main Process Supplier during the Project Implementation Stages

Task	Owner/User	Main Contractor/Main Process Supplier
Effective Contract	Main task holder	Might also be involved
Government Approvals	Part task holder	Part task holder
Engineering	Approval	Main task holder
Engineering Approval	Approval	Main task holder
Test of equipment & material	Approval	Main task holder
Site ready for construction	Main task holder	Approval
Site Mobilization	Approval	Main task holder
Quality of installation	Approval	Main task holder
Progress Measurement	Approval	Main task holder
Installation	Approval	Main task holder
Pre-commissioning check out	Approval	Main task holder
Raw material fulfils spec.	Main task holder	Approval
Commissioning & Training	Approval	Main task holder
Test run	Approval	Main task holder
Acceptance	Approval	Main task holder
Warranty	Main task holder	Main task holder
Operation & maintenance	Main task holder	Approval

During the project implementation stages this relation will normally come under heavy stress due to the number and seriousness of issues to be dealt with and agreed between the parties. This stress can cause disagreements to become conflicts. Therefore all what reasonably can be done to prepare and minimize this stress handling will be beneficial from a conflict prevention point of view and from an overall project result point of view! Most projects have "room for improvement" re-

Fig. 2.3. Grain terminal and storage plant in Eastern Europe

garding the distribution of scope of work, definitions of limits of supply and work, equipment specifications and functional specifications at an early stage. These areas cause most of the disagreements and of the conflicts in international projects. Efforts to reduce these causes before contract drafting and negotiations are recommended.

Another increasingly important relation is the one between the Owner/User and the Main Process Supplier/Main Contractor on one side and the local Authorities on the other. Regulatory matter is often becoming a real critical project issue symbolized by the abbreviation QHSE (Quality Health Safety Environment). This aspect is further discussed in 2.5 below.

2.3 Analysis of Contracting Parties and Their Interests

We recommend that each contract partner spends some time on mapping and analyzing information about the other party and its key representatives before project implementation starts. The effort and costs involved are marginal compared to what is spent once a major conflict has broken out, and it helps preventing such break outs! The checklist in Schedule 2.1 gives the main points of such an analysis.

2.4 Conflict Contingency Plan

The Project Plan or the project implementation manual – that the Project Manager and his team update or prepare for the project start-up meeting – should contain a chapter regarding conflict prevention and resolution including procedures and instructions that all contractors' staff involved should follow. This conflict contingency plan can be built-up on the basis of the following main points:

1. Careful reading of the official "Project Description" in order to understand the project objectives, scope and parties
2. Very careful reading of the "Contract with all its appendices" in order to be familiar with the parties' responsibilities and the contractual procedures
3. Understanding and interpretation of the contracting structure and the contractual procedures for the relations between the parties (contractual procedures are the descriptions of the dynamic interaction between the contract parties e.g. design approval procedure or variation of scope of work procedure etc.). Company rules of delegated authority to be respected.

2.5 Laws, Regulations, Decrees and Standards Governing the Contract

The project and the work to implement it (especially the work carried out in the project country) are subject to the Laws of the project country. The Laws include

Schedule 2.1. Checklist for analysis of project parties and their interests

A. Standard company information as annual accounts and registration information
   1. Group level information
   2. Company level information
   3. Business Unit information
   4. Department information
B. Standard company information required
   1. Annual report and company organization chart
   2. Company and Management registration and CV's
   3. Credit information (credit-worthiness)
   4. Reference list – projects, clients and subcontractors/subsuppliers
C. Key Persons
   1. Senior Management level
   2. Management level including Project Management
   3. Key professionals
   4. Advisers working closely with one of the parties
D. Areas of special attention
   1. Financial situation generally
   2. Annual Accounts from last 3 years and Accounting practices
   3. Project references
   4. Personal & Professional references and personal contacts
E. The Role of the Consultant and the Inspection Authority
   1. Stipulations in the contract
   2. Professional competences
   3. Terms of reference
   4. Duration of assignment
F. Relations between Main Contractor and the subcontractors
   1. Type of subcontract
   2. Back-to-back stipulations
   3. Procedures involving subcontractors
G. FIDIC Contract Conditions: The Engineer
   1. Stipulations in the contract
   2. Professional competence
   3. Terms of reference
   4. Duration of assignment

Regulations, Decrees and Standards issued by the Authorities of the project country. Reference is made to Chap. 3 discussing taxes and duties and Chap. 8 dealing with local Technical Decrees and Standards.

In case of any contradiction between the Contracts of the project and the Laws etc. of the project country the latter has priority. If a solution can be found that fulfils both, it will generally be the one preferred by the bureaucrats, but not necessarily by the contractor who often has to bear the extra costs. Here is a potential conflict area.

Especially turn-key contractors have to investigate and familiarize themselves with the Laws, Regulations, Decrees and Standards governing the Contract, because turn-key contracting can be compared to setting up a local company and liquidating it again after project completion.

Local income tax of expatriate staff working in the project country is becoming an increasing focus area by local tax authorities! And this also goes for local Value Added Tax (VAT) and company profit taxes etc.!

Extra project costs due to insufficient planning and preparations in the area of local Law and Authorities are generally among the big contributors to project budget overrun!

2.6 The Authorities of the Project Country

Basically the project needs three types of approvals from the Authorities of the project country:

A. Approvals to establish, build and operate the project prior to start of engineering and construction – this involves the following investigations and approvals:
   - Basic process design and plant lay-out necessary for the approvals
   - Soil investigation
   - Plant location approval
   - Infrastructure and utilities approval
   - Environmental impact assessment
   - Environmental approval
   - Construction permit
   - Industrial production license
   - Import license for equipment etc.

B. Permit to operate the plant as built correctly in accordance with the approvals prior to project start:
   - Approval of Safety & Health instructions as part of the Operational Manual and of the Maintenance Manual
   - Approval of environmental protection measures, controls and instructions
   - Approvals of fire prevention documentation and training, fire fighting procedures and training as well as safety instructions in case of explosions

C. Plant operation compliance with the project approvals:
   - Safety, Health & Environmental instructions and proper plant documentation, training and compliance inspections
   - Fire prevention, fire fighting and explosions safety instructions and proper plant documentation, training and compliance inspections

It is our advice to prepare and perform the necessary investigations, engineering and planning up front and if necessary use of external professional services as a supplement to one's own internal preparations (see Table 2.4). The Owner should be main task holder and overall responsible for this activity. The Owner can delegate work to consultants, so can process suppliers and contractors but the overall responsibility remains with the Owner, Contractor and Process Supplier!

Table 2.4. Recommendations regarding handling of authorities in the project country

No. of the activity	Activity recommendation	Remarks and notes
1.	Pre bid desk study of relevant Laws	Necessary but not sufficient
2.	Pre contract field study of relevant Laws, Regulations and Standards	Very critical – expert assistance needed in most cases
3.	Establishing local set up to deal with Authorities	
4.	Establishing procedures for import of equipment and materials	
5.	Establishing procedures for work permits	
6.	Establishing procedures for accounting, declaration and payment of local taxes	
7.	General recommendations	Play it the bureaucratic way; Documentation as required; Contractor should take initiative

2.7 Other Players

A. Local banks
B. Non Governmental Organizations (NGO's)
C. International banks
D. International Finance Institutions (IFI's) as International Finance Corporation (IFC), African Development Bank, Deutsche Entwicklungs Gesellschaft (DEG) etc.
E. Foreign Governmental Aid Organizations

The banks, IFI's and aid organizations might indirectly be contract partners via the finance agreement. Reference is made to Felding et al. (2002)

2.8 Conclusion

Conflict between the parties can be prevented or handled if there is a better knowledge and understanding of the other's situation and interests.

A careful distribution of the responsibilities and duties in the project between the parties must be made meticulously and documented already in the project development stage. It must be based on the project's total requirements and on the parties' capabilities and not on wishful thinking or cost hiding exercises. It must also take the regulatory requirements into consideration. These requirements increase the complexity of projects and lead to a higher risk of disagreements and conflicts. The countermeasures are: preparations, dialogue, more preparations and more dialogue and finally agreements between the parties!

More emphasis on these aspects before and during preparation and implementation of contract drafting and negotiations is recommended as described in the next Chap. 3 and will lead to better contracts and thereby reducing the risk of conflicts.

References

Turner J R (1993) The handbook of Project-based Management. London
Mikkelsen H & Riis, JO (2004) Grundbog i Projektledelse (Basics of PM). Prodevo, Copenhagen
Felding F, Rindorf E, Sejersen E (2002) Eksport- og Projektfinansiering. Published by the Authors in Copenhagen

3 Engineering, Supply and Construction Contracts

Abstract. Project conflicts are often caused by the parties' differing interpretation of the contract provisions. Some provisions deal with the obligations of the parties, and if the specifications are ambiguous they give rise to disagreements and conflicts. Other provisions deal with procedures for approvals, plant acceptance, etc. and, if they are impractical or unclear, conflicts may arise. Finally there are contract provisions about how to resolve conflicts, mostly relevant when a conflict has materialized.

Contracts are binding for the parties (pacta sunt servanda, as the Romans said) once signed and enforced e.g. by down payment received etc. An effective contract cannot be cancelled without major compensation to the other party, to the extent that this way out is generally not feasible. The contract execution normally ends by commercial use and signing the acceptance certificate etc. The warranty period then starts but it normally does not cause major problems.

With respect to conflict handling the contracts of most interest are the ones, where "time is of the essence". Time is by far the most important factor for the success of a project. Combined with a fixed price, advanced technology and fast track schedules these contracts simply do not allow for conflicts, defined as serious disagreements, hampering the daily project cooperation and progress.

Fig. 3.1. Interior of a Seed Processing Plant in Eastern Europe

Fig. 3.2. Exterior of a Seed Processing Plant in Eastern Europe

Most Engineering, Supply and Construction project conflicts are related to the interpretation and the use of the Contract. In our opinion the number and the extent of project conflicts can be reduced by improving the contracts in terms of clearer definitions, specifications, procedures etc. and therefore this handbook deals with how to improve a Contract.

Key words: Engineering; Supply; Construction; Contracts; Equipment Supply Contracts; Turn-key Contracts; "time is of the essence"; "fit for the purpose"; "fixed price"; "agreed time"; material obligations; financial obligations; owners obligations; project organisation; dispute resolution; scope of work; specifications (or specs); transfer of risk; "Contractor's All Risk Insurance (CAR insurance)"; INCOTERMS; International Chamber of Commerce (ICC); Force Majeure; Regulatory obligations; Quality; Health; Safety; Environment; risks; delay; FIDIC; ORGALIME; standards; general conditions; special conditions; tender documents

3.1 Contract Types and Their Elements

Combined engineering, supply and construction (or construction supervision) contracts can be subdivided into three categories:

- Equipment supply contracts including or combined with supervision contracts – defined as supply of specified equipment without any plant or process responsibility other than the performance of each individual piece of equipment
- Complete project supply contracts including or combined with supervision (of installation and commissioning for which the Owner is responsible) contracts – defined as supply of a process line or a plant including the performance of the line or the plant (process guarantees and the general requirements: "fit for purpose")
- Turn-key contracts that are characterised by the supplier-contractor's full responsibility for equipment, erection, documentation, installation and commissioning,

for the functioning of the plant or process line within an agreed time frame and for an agreed fixed price.

It is obvious that there are grey areas between the three groups and generally grey areas are dangerous as they tend to fall into a category of higher risk, contrary to what might be expected.

The turn-key concept is characterised by the following 4 factors and they are to some extent also present in the complete supply contract:

1. Completeness of engineering, supply and documentation according to the "fit for the purpose" principle (overrules equipment spec. etc.) unless otherwise specified.
2. Fixed price that can only be revised by variation orders signed by the Client (Buyer, Employer, Owner).
3. Agreed time frame with a firm delivery or completion date – that can only be changed by time extension in a contract amendment – as contractor's responsibility. If not kept, the contractor has to compensate the owner for damages.
4. Performance guarantees which have to be demonstrated and fully achieved, if compensations from the Contractor to the Client should be avoided.

Combined engineering, supply and construction contracts consist of the following elements:

- Material obligations e.g. scope of supply, specifications, client supply, changes and other similar provisions.
- Regulatory obligations e.g. construction permit, environmental approval etc.
- Financial obligations e.g. payments, prices, conditions & delivery terms and financial guaranties etc.
- Taxes, i.e. which party is responsible for the declaration and the payment of each type of taxes and duties applicable.
- Project organisation provisions, e.g. project and site manager's duties, project meetings, communication etc.
- Dispute resolution provisions e.g. choice of law, mediation, arbitration etc.

3.2 Material Obligations and the Drafting Process

The Contractor's scope of work, the technical specifications and quality requirements are key issues during contract negotiations. Despite the attention paid and the drafting effort made, the specifications and quality requirements often cause conflicts during project implementation in the case of design and equipment approvals, training of personnel, documentation, installation and commissioning.

Potential conflicts might be avoided by focussing on three factors during the contract drafting phase:

A. The close and constructive dialogue between contractor and owner is essential in order to achieve a comprehensive description of the scope of work and the specifications of the supply and works. The supplier or contractor is the specialist for the process, for the equipment, for the utilities and for the buildings, and the Client or Owner has the operational and commercial experience. Consensus and clarity should be obtained between them. It is strongly suggested to keep all documents exchanged between the parties during contract negotiation, in order to prove, in case of litigation, the intention of the parties.

B. The scope of work and price relation is a key negotiating aspect during selection of the contractor and negotiation of the Contract with him. This aspect unfortunately works against the dialogue and cooperation mentioned above under point A. But still a lot can be done in order to "defuse" the tensions and reach clear and operational specifications. Contractors in our opinion have to be courageous in this respect.

C. The development and design of a technical solution and its scope of work and specifications for the project is very challenging as time is short for the project development. At the latest, when the final design of buildings and foundations and the ordering of equipment have to be done, the scope and specifications have to be ready, clear and final. This dynamic aspect is not normally reflected in the contract and is against the principle of fixed price and firm time contracts. The new partnership approach might be a way out of the dilemma – experience will show. But anyhow improvements are definitely necessary!

The list below contains the typical contract documents that determine the scope and specifications of the technical solution. It illustrates, that developing and drafting of the specifications are a major task:

- Process and control system descriptions and diagrams
- Electrical single line diagram
- Plant description and lay-out drawings
- Descriptions and specifications of the plant site, soil conditions, foundations and buildings
- Performance and process guarantees and test procedures. Specification of raw materials and utilities necessary to achieve performance guarantees
- List of machinery, electrical equipment and instrumentation with their specifications
- Interface of supply/work between owner and contractor or between supplier and subcontractor
- Quality Health, Safety and Environmental requirements. Regulations and Standards applicable
- Procedures and requirements for approval of design, engineering and installation
- Time schedule as bar chart and/or as PERT-diagram (see Chap. 8)
- Requirements regarding testing, inspection, pre-commissioning, commissioning and performance/acceptance tests
- Contract provisions regarding inspection, training, variation orders, management of milestones, planning, progress measuring and reporting
- Contract provisions regarding documentation of the plant design, construction, operation and maintenance
- List of required and furnished spare parts

In turn-key contracts it is especially important to clearly specify the Client's scope of work and responsibilities as they constitute exemptions to the main rule that all supplies, works and services are the responsibility of the Contractor. Often these exemptions arise because the Owner/Client wants to supply, pay and take responsibility for certain supplies and thereby reduce the turn-key price.

The clear interface of supply between Contractor and Client are very important for a productive cooperation between the two parties during project implementation.

A typical list of Client's supply and services could contain some of the following elements:

- Permits and licenses from the authorities
- Soil tests of plant site, raw material availability and quality surveys
- Import licenses, customs clearance and local transport from arrival port to construction site
- Site preparations for construction work incl. facilities and access road
- Utilities such as communication systems, water, sewage, electricity and gas
- Civil works incl. foundations, buildings with their internal completion, internal roads etc.
- Supply of certain equipment, material and services as a supplement to the Supplier's/Contractor's scope of supply
- Installation and testing of the equipment under the supervision of the supplier
- Raw material, utilities and operational staff for commissioning activities
- Accommodation, canteen, local transportation and facilities for site staff and supervisors
- Handling of local taxes and custom's duties and payments.

In normal supply contracts the risk transfer for the equipment is quite simple. The supplier is only responsible until shipment and the equipment is taken over by the freight forwarder. Thereafter the supplier has the responsibility for a professional supervision during installation and commissioning.

The transfer of the risk from the Contractor to the Owner should not be confused with the transfer of ownership that in turn-key contracts typically takes place earlier than the risk transfer. Therefore the distribution of risks during project implementation is quite different for turn-key contracts than for supply contracts. In turn-key contracts the risk is transferred from Contractor to Owner at the latter's Acceptance. This has an influence on the insurance coverage. Contractor's All Risk (CAR) insurance is absolutely necessary in turn-key contracts or hybrids of turn-key and plant supply contracts.

For an equipment supplier who gradually moves towards the turn-key type of contract with more extensive responsibilities the "grey area" between normal supply contract – where the supplier delivers only the specified equipment according to INCOTERMS FCA or CFR on one hand – and a full fledged turn-key contract (even excl. civil works) on the other side – is the most dangerous area. It is often understood as an extended supply contract and in fact it is more turn-key like. INCOTERMS (INternational COmmercial TERMS is an international standard term of sale) are issued

Schedule 3.1. The 13 INCOTERMS in short

A. Origin Terms
   - Ex Works (EXW), named place where shipment is available to the Buyer, not loaded. The Seller will not contract any transportation.

B. International Carriage NOT paid by Seller
   - Free Carrier At (FCA), unloaded at the seller's dock or a named place where shipment is available to the international carrier or agent, not loaded. This term can be used for any mode of transportation.
   - Free Alongside Ship (FAS), named ocean port of shipment. Ocean shipments which are not containerized.
   - Free On Board vessel (FOB), named ocean port of shipment. This term is used for ocean shipment only where it is important that the goods pass the ship's rail. Before this term was the most used in this group but it will gradually be taken over by FCA.

C. International Carriage paid by Seller
   - Cost and FReight (CFR), named ocean port of destination. Ocean shipments which are not containerized.
   - Cost, Insurance and Freight (CIF), named ocean port of destination. Ocean shipments which are not containerized.
   - Carriage Paid To (CPT), named place or port of destination. Air or ocean shipments which are containerized and roll-on roll off shipments.
   - Carriage and Insurance Paid to (CIP), named place or port of destination. Air or ocean shipments which are containerized and roll-on roll-off shipments.

D. Arrival at Stated Destination
   - Delivered At Frontier (DAF), named place of destination, by land, not unloaded. This term is used for any mode of transportation but must be delivered by land.
   - Delivered Ex-Ship (DES), named port of destination, not unloaded. This term is used for ocean shipments only.
   - Delivered Ex-Quay (DEQ), named port of destination, unloaded and not cleared. This term is used for ocean shipments only.
   - Delivered Duty Unpaid (DDU), named place of destination, not unloaded, and not cleared. This term is used for any mode of transportation.
   - Delivered Duty Paid (DDP), named place of destination, not unloaded but cleared. This term is used for any mode of transportation.

by the International Chamber of Commerce. Reference is made to Schedule 3.1 "The 13 INCOTERMS in short".

The so called turn-key aspects, which require special attention, are the following:

- Responsibility for a complete process line (incl. necessary non specified equipment) and responsibility for process performance
- Responsibility for installation and commissioning at a firm time and fixed price incl. local workforce (besides the design, supply and documentation)
- Responsibility for local factors, such as taxes/duties, transport from arriving port to plant site, raw materials and utilities
- Fit for purpose contract provision ("as required for the proper design, execution, operation and maintenance of the plant in all respects")

Schedule 3.2. Checklist for Main Points in Project Contracts

(the list of sub-points is not exhaustive, it includes only the critical points requiring special attention)

1. General conditions
   - Standard as ORGALIME or other
   - Tailor-made for the project
2. Scope of supply and work (conflict prevention critical area)
   - Scope & specification of design, equipment, documentation and training etc.
   - Scope & specification of client supply
   - Battery limits and borderline activities to be clearly specified
   - Erection and commissioning: Execution or only supervision
3. Performance (or functional) guarantees (conflict prevention critical area)
   - All performance guarantees in one contract section
   - Performance guarantees based on "input guarantees"
   - Design parameters are not performance guarantees unless specified
   - Liquidated damage paragraph attached to each performance guarantee
4. Price and terms of payment
   - Total contract value and price break down
   - Shipping, progress measurement and invoicing (agreed form in annex)
   - Payment terms and conditions (form to be agreed in annex)
   - Letter of credit or bank guarantee (form to be agreed in annex)
5. Time of delivery and liquidated damages (conflict prevention critical area)
   - Project activity plan and time schedule (form to be agreed in annex)
   - Delivery terms as INCOTERMS and inspection for supply
   - Delivery time and place
   - Planning procedures (form to be agreed in annex)
   - Liquidated damages and related conditions
6. Commissioning and acceptance (conflict prevention critical area)
   - Testing in workshops or on site prior to erection (form to be agreed in annex)
   - Commissioning tests: conditions, procedures, planning and approval of results
   - Other acceptance inspections and approvals
   - Acceptance certificate and start of warranty period
7. Client's financial guarantees
   - Advance payment bond (agreed form in annex)
   - Performance bond (agreed form in annex)
   - Warranty bond (agreed form in annex)
8. Risks
   - Risk transfer & exempted risks
   - Insurances
   - Force Majeure
9. Settlement and litigation (conflict prevention critical area)
   - Technical expertise & mediation
   - Court proceedings or Arbitration
   - Choice of place and selection of Law
10. Annexes (conflict prevention critical areas – see Schedule 3.3 & 3.4 below)

Turn-key contracts normally contain a clause regarding exempted risks – exemptions from the main rule that the risk is with the main contractor until acceptance certificate. Exempted risks could be terrorism, war, civil riots, earthquake and flooding etc, normally defined as "Force Majeure". These are risks, which the supplier or contractor can not be expected to cover. The "Force Majeure" clause means that the contractor might get time extension but no compensation for the extra cost of a possible delay or a catching up.

3.3 Regulatory Obligations

The regulatory aspects and obligations related to major projects and their contracts have developed dramatically in the last 30 years, mainly due to two reasons:

- New and more comprehensive legislation regarding Quality, Health, Safety and Environment
- The trend to outsource project implementation in turn-key type of contracts

Before major international contracts contained very few regulatory provisions e.g. building permit, local taxes and import license were declared as Client obligations.

Schedule 3.3. Checklist for the Technical Specifications of a Turn Key Contract, Part 1

A. Scope of work and limits of supply and works
   - Scope of work for civil, building and steel structure – (contractor's scope, owner's scope and their limits are specified by border lines on drawings, border line activity schedule and description on key locations)
   - Scope of work mechanical equipment (supply and installation/testing)
   - Scope of work electrical equipment (supply and installation/testing)
   - Scope of work instrumentation and automation (supply and installation/testing)
   - Scope of work training
   - Scope of work of commissioning

B. Technical specifications
   - Specifications of civil, building and steel structure works
   - Specifications of mechanical equipment (supply and installation/testing)
   - Specifications of electrical equipment
   - Specifications of instrumentation and automation

C. Plant performance
   - Process description in words and diagrams
   - Plant performance guarantee (output)
   - Plant performance guarantee conditions (input)
   - Performance test procedure
   - Performance test result, evaluation and approval

D. Planning and scheduling
   - Project time schedule as per contract
   - Planning procedure
   - Delays

E. Drawings and documentation
   - Planning procedure

Typical supplier obligations were "country of origin certificate" and "equipment test certificate" according to export country technical standards and export country taxes and permits.

This has changed radically and the main supplier/main contractor now becomes much more involved in taking responsibility for obtaining construction and environmental permits etc. The following tasks are now often under the responsibility of the contractor:

- Approvals and permits before the site activities are allowed to start
- Complaints handling
- Inspections during implementation
- Documentation for verification by authorized professionals
- Final as built approval as a condition for operation

All this takes time, requires resources and must be planned carefully. The consequences in case of problems can be quite serious:

- Delays in construction start-up and in the duration of the execution
- Extra manpower and external consultants to deal with the regulatory problems
- Extra cost regarding documentation and verification

In order to reduce the risk of conflicts, the distribution of responsibility for regulatory matters must be clarified. The distribution of responsibility for regulatory matters has become critical for the success of project implementation and has to be dealt with in a comprehensive manner before the contract is signed:

- It is recommended to carefully investigate the regulatory requirements at an early stage – this investigation should not be under estimated and both parties must have the interest in this clarification!
- Clear distribution of regulatory obligations to be performed either by the Client or by the Contractor based on the work breakdown structure (see Chap. 8)
- Possible delay consequences and time extension as well as extra resource requirements and cost consequences should be considered if regulatory problems arise
- We recommend facing realities as to dividing the responsibility for regulatory issues between the parties. The party with the best ability and the party whose work is most dependent on the outcome should be chosen to be responsible. It is obvious that without any provisions to the contrary the Main Contractor is responsible for most of the regulatory issues in a turn-key contract as well
- This means that a section in the legal part of the contract and a number of annexes should deal with regulatory issues. The old provision of "the Client shall assist the Contractor to obtain import license and building permit" has gone!

3.4 Financial Obligations

3.4.1 Price and Payment Terms

For obvious reasons prices and payment conditions are subject to difficult contract negotiations, since they influence such vital points as costs, liquidity, profit and security. During contract execution the two contracting parties have opposite cash flow

Schedule 3.4. Checklist for the Technical Specifications of a Turn Key Contract, Part 2

A. Staffing, site regulations and site facilities
   - Staffing
   - Site regulations
   - Site facilities and border line activity schedule
B. Design meetings and design approvals
   - Design meetings: purpose, preparation, agenda and minutes of meeting
   - Design approval procedure and rules
C. Acceptance, taking over and warranty period
   - Acceptance procedure and agreed forms
   - Taking over consequences
   - Suppliers' warranty obligation
   - Warranty claim procedure
D. Quality, Health, Safety, Environment (QHSE)
   - Quality procedure and rules
   - Health procedure and rules
   - Safety procedure and rules
   - Environmental procedure and rules
E. Project and Site Management, project meetings and variation orders
   - Project and Site Management – approvals, powers, instructions
   - Project meetings – agenda and frequency
   - Variation orders – procedure and forms
F. Customs clearance, Taxes & Duties, Insurances, Invoicing, Payments, Financial guarantees
   - Customs clearance – procedure and template
   - Taxes & Duties – procedure and forms
   - Insurances – scope of insurances, insurance companies and claims procedure
   - Shipping, Invoicing and Payments
   - Financial guarantees

interests and the conditions are vital for both of them because liquidity is often under pressure.

The party in possession of the cash has a dominating position. It is the other party who wants the money and who has to produce the evidence and documentation enabling the first party to make the payment or release the securities. Only if the documentation is as agreed, then the first party has an undisputable obligation to make the payment etc. Even if the contractual relation is characterised by cooperation the "power play" aspect is still present behind the scene. Under stressed conditions the "power play" aspect might be more dominant and harmful to solution finding.

The main factors in the price and payment conditions for Engineering, Supply and Construction contracts are the following:

- Price specifications
- Payment terms, conditions and instalments
- Payment securities
- Financial risks (transfer of payments, payment delays, exchange rates, payments in local currency etc.)

- The relation between the conditions of the Engineering, Supply and Construction contract and the Financing Agreement (Banks and International Finance Institutions will have requirements to the Engineering, Supply and Construction Contract and might have to approve it)

Generally those provisions are seldom causing conflicts, because they are very carefully drafted and agreed between the parties and approved by the banks and the International Finance Institutions.

3.4.2 Security System

- On order of the Client the Bank (or insurance company) will issue the financial guarantees as security for the client's payment obligations
- On order of the Contractor the Bank (or insurance company) will issue the financial guarantees as security for the contractor's performance obligations (plant performance, function, time, compliance and quality)

The bank which issues a financial guarantee does not become party in the supply or turn-key contract. The bank guarantees the payment of a certain amount, should the Contractor not be able to perform according to the contract. The bank will take recourse to the Client or Contractor, as the case might be, and on whose request the guarantee has been issued.

Financial guarantee types:

- Conditional guarantees or bonds (can only be called/paid out upon confirmation by a third party or upon a legal award)
- On demand guarantees or bonds (can be called/paid out on simple written demand by the other party)

In the Project Contracts the following 4 guarantee types are most common:

- Bid Bond or Tender Guarantee
- Advance Payment Bond, Prepayment Bond or Down Payment Guarantee
- Performance Bond or Completion Guarantee
- Retention Bond or Warranty Guarantee

The widespread use of "on demand guarantees or bonds" and the practise of calling these guarantees in case of a conflict are matters of concern, because they tend to increase the conflict. We recommend, that the contract parties carefully consider calling a guarantee and that they first try to imply the conflict resolution methods of Sect. 6.3, before taking such a drastic step, which could prove unnecessary and could cause obstruction.

3.4.3 Relation Between the Engineering, Supply and Construction Contract and the Project Finance Agreement

- The Client will make a Finance Agreement with the Bank (or Financial Institution) and the Client and the Contractor will make such the Contract as well. The agreements are separate and different. Nevertheless certain provisions and factual aspects are the same (e.g. Contractor payments to be drawn on the loan) but legally they are absolutely independent of each other.

- The Engineering, Supply and Construction contract and the Finance Agreement have to be operative independently but each one will only become effective if the other one is signed and effective.
- The payment conditions in the contract have to be coordinated carefully with the disbursing conditions of the finance agreement. All terms and conditions must be compatible with similar terms and conditions in the Engineering, Supply and Construction Contract.
- The Financial Institution does not at all want to be involved legally in the Supply or Construction Contract between the Client and the Contractor. This separation safeguards the Financial Institution from being drawn into any conflict resulting from the Engineering, Supply and Construction Contract. Therefore the bank requires the Contractor to be liable for any payments drawn on the loan in the form of legal recourse.

The Financial Institution must, as required, be informed about the project progress and risks by the Owner and/or the Contractor in the form of status and progress reports.

3.5 Taxes

Taxes and duties related to the project implementation and especially those payable in the project country are of increasing and presently underrated importance. Governments show an ever increasing creativity regarding new and complicated taxes and related legislation and administration. Therefore the turn-key contract operator should pay attention to taxes in the project country. The main issue is who is going to declare and pay the taxes during project implementation and who is ultimately bearing the cost and how?

The old equipment supply provision in the proposal or in the contract: "the Buyer is responsible for taxes caused by the project in the project country and Seller in countries outside the project country" is not sufficient any longer and if not replaced by a new and comprehensive provision the total tax bill risks ending up with the supplier/contractor. General provisions that the Buyer will reimburse the Seller (or Contractor) the local taxes, he has to pay in the project country, are not at all safe, because of likely disagreements on how to minimize taxes by various measures. Reimbursement is difficult to manage because the party, who is reimbursing and ultimately bearing the cost, can claim, that the party declaring and paying in the first instance did not do it the right way and therefore the tax burden was heavier than necessary and that part of it will not be reimbursed!

In certain countries the local tax risk might be very substantial, and as the most important project risk it can in some cases amount to approximately 5–10% of the contract value. Although this level is only applicable in certain countries, local taxes represent an important risk and administrative challenge in most of the countries.

The taxes and duties related to the project implementation can be of the following types:

- Import duties and levies (incl. harbour costs, custom clearance etc.)
- Income taxes for staff and workers (incl. expatriates)
- Company and profit taxes and duties
- Trading taxes as Value Added Tax or Sales Tax
- Vehicle taxes and duties
- Other local duties and taxes (e.g. water and waste)

When a contractor or supplier enters the market in a new country, he needs to investigate the local tax situation carefully in order to insert the right costs in his estimation, the right man-hours in his project staffing and the right conditions in his proposal. Before a new project starts, handling of local taxes should be one of the top priorities on the action list for mobilisation. This includes providing for internal and external resources and appointment of advisers to perform the necessary preparation and registration. Otherwise the tax situation might turn into a "nightmare" distracting the project management's attention from project progress and quality, and cause project conflicts.

3.6 Project Organisation Provisions

The contract should contain provisions for the following elements:

- Name and authority of the Representatives of Client and of Contractor, as well as their deputies. This type of provision is becoming quite normal in complete plant supply and turn-key contracts, and it is very useful.
- Project meetings and progress reporting: These instruments are given too low priority in contract drafting and too little emphasis in project implementation. Used professionally, this is one of the most important conflict prevention measures.
- Engineering and equipment approvals by Client (person responsible, procedure, format, information requirements, deadlines, recourse etc.). Reference is made to Table 3.1.

3.7 Dispute Resolution Provisions

Reference is made to Chap. 6 "Litigation, Arbitration and Mediation Contributing to Conflict Settlement".

Table 3.1. Main Contract Functions and Related Recommendations

Contract function	Drafting recommendation	Application recommendation
Making the contract operative	Specify the order, content and documentation of fulfilling each condition and step.	Follow the contract meticulously and ensure the other party's approval of each step.
Approval of engineering and of equipment procurement	Specify a workable procedure with clear rules in case of disagreement. Adjust the responsibility level to the resources available.	Start early and have enough resources to handle approvals in order not to delay the process and the project.
Variation orders	Specify a simple and workable procedure with clear rules in case of disagreement.	Try out the procedure early in the project with minor variations and get it to work by professionalism and flexibility.
Activities planning and scheduling	Specify a simple and workable procedure with clear rules in case of disagreement.	Try out the procedure early in the project with minor coordination issues and get it to work by trust and flexibility.
Construction approvals	Concentrate on major issues. Specify a simple and workable procedure for construction approvals with clear rules and fast reaction.	Concentrate on major issues. Make a major effort to get it to work in practise right from the beginning and keep it that way – it is needed at the end!
Commissioning approvals	Specify a simple and workable procedure for commissioning approvals with clear rules in case of disagreement.	Make comprehensive draft plan, content and procedure early in the project, discuss it with the Owner and agree. Follow the agreement.
Plant acceptance	Specify the order, content and documentation for each condition and step in plant acceptance.	Follow the contract meticulously and ensure the other party's approval of each step. Agree on Acceptance Certificate form prior to commissioning.
Warranty claim handling and settlement	Specify a simple and workable procedure for warranty claim handling with clear rules in case of disagreement.	Try out the procedure before acceptance with a minor warranty case and get it to work by trust and flexibility. Give and take attitude might help!

3.8 Contract Drafting in Europe, Asia and America

3.8.1 Introduction

Despite the fact that every contract drafting situation is different, every Owner and Contractor are different and every industry and country are different, there seem to be some general tendencies, which are similar in Asia, America and Europe. In this section we want to discuss such general tendencies – despite the existing uncertainties – in order to start a discussion among professional project people; a discussion,

Fig. 3.3. LPG (Liquefied Petroleum Gas) Plant In Africa

which hopefully will lead to a better understanding of how to build up good relations between the parties in a contract drafting situation. This might eventually lead to better contracts and to fewer disputes during execution of the contract. This will be of benefit to both parties.

Fig. 3.4. Filling Hall of LPG Plant In Africa

3.8.2 Survey on Lawyers' Opinion on Conflicts in International Projects

We have performed a mini survey among lawyers working with international project contracts and asked the following questions to lawyers in USA, France, Germany and Denmark.

We summarize their answers as follows:

Question 1. What is the tendency in number and seriousness of contractual conflicts as you see the trend?

Generally the Lawyers agree on an increased risk of conflicts in international projects especially, where major issues with respect to financial consequences or Industrial Property issues are involved. The tendency that disagreements more often result in disputes and less often in settlements is mentioned.

One answer mentions that if major financial consequences are at stake the contracting parties have a tendency to try very hard to maintain and defend their position and carry out the dispute up to a point, where they have to realise that it will take several years of work, intensive, time consuming and expensive court proceedings in order to get a final decision in the matter. It can very often be observed that the parties have to reach a point where they realise these facts before they are willing to decide, that it is more reasonable to settle the dispute. Until then, however, the parties already spent a considerable amount of money to prepare their own claims and to defend themselves against the claims of the other party. (The Authors add that then it might be too late to negotiate!)

This tendency of pushing a dispute to the limit is due to the fact that managing directors and project managers are under a growing pressure in our global economic world in terms of time and profit. They feel they have to do their utmost to make the project as profitable as possible which often prevents them from giving in on certain unfavourable positions. Another situation where parties push a dispute to the limit is, if the parties are emotionally extremely involved, so that the dispute becomes a matter of principle. This especially applies to cases where the contracting parties do not know each other well and where there is not a lot of experience in cooperating with each other.

Question 2. What are in your opinion typical contractual conflicts?

The Lawyers mention the following common areas of disagreements:

a. Disagreement regarding delays, in particular delays of the contractor.
b. Disagreement regarding the scope of work, specification and functionality. There are very often conflicts regarding the technical specifications which are not well elaborated and defined. Further, contracting parties regularly disagree on what is to be understood as the scope of the contract and what, by contrast, is to be considered as extra work of the contractor for which he can claim additional payment. Mostly this is due to a lack of defining and/or adhering to specific rules of change requests of the client or with unforeseen difficulties during the performance of the contract.

c. Multiple type conflicts. This includes the disagreements mentioned above, plus special problems on the part of one party such as financial problems or poor project management.

The most important causes for conflicts mentioned in the Lawyers' replies are the following:

- Time pressure due to a far too ambitious time schedule for the project, which does not leave any space for unforeseen circumstances or problems.
- Poor contract drafting, in particular with respect to procedural rules (e.g. change requests, extra work, the performance to achieve, the scope of work or technical specifications). All this leads to misunderstandings, misinterpretations of the contract and to misjudgement of one's own rights and claims.
- Poor project management on part of the contractor or on part of both parties.
- Lack of ability or willingness to cooperate with the other party: This happens quite often, in particular if the parties cooperate for the first time or if there is a lack of understanding of the other party‘s position, sometimes due to cultural differences. In most cases such conflicts are caused by a lack of communication between the parties.
- Weak financial situation of one or both contract parties: Does not occur very often. If a financial crisis occurs during a project, it is very often one of the sub-contractors of the contractor who suffers a lack of liquidity due to outstanding payments. Of course, this has very often a major impact on the project depending on the scope and importance of the sub-contractor's work.
- Misinterpretation of the contract with respect to time schedule, scope of work, specification of performance or specification of equipment.
- Misjudgement of one's own claims.
- Lack of ability or willingness to cooperate with the other party. The lack of co-operation is often found as a lack of understanding of the other party's situation with respect to the contract and to cultural differences.

Question 3. What is the trend in the parties' ability to prevent contractual conflicts as per your experience?

In their replies the Lawyers mention the following trends:

- Awareness of the cost involved – both in terms of financial figures and the drainage it puts on an organisation will in general create a big focus on settling the matters. However, less people seem to know exactly how to go about this (i.e. settling by negotiations between the parties), – a discipline which needs to be learned – and adapted from time to time, depending on the matter at hand.
- There seems to be a tendency that the contractors take their chances in arbitration or in court, unless a settlement can be reached on very favourable terms.
- There is a tendency to push an existing dispute to the limit and to maintain one's position in order to achieve a favourable result. The ability and willingness to solve a dispute in an amicable way is in general higher, if the contracting parties have a long-standing business relationship and are used to cooperate with each

other. This can be observed in major shipbuilding projects where very often the same companies are involved in several projects and therefore the persons acting on behalf of the companies know each other very well. In such projects the parties usually try to solve the conflicts internally and as quickly as possible.

Question 4. What is the trend in the parties' ability to solve contractual conflicts by negotiation between them as you see it?

In their replies the Lawyers mention the following trends:

- Generally the answers indicate that the negotiated settlement is under pressure and less wanted compared to litigation or arbitration.
- But in many cases there seems to be an understanding between the parties that disputes arise in international contracting and that such disputes need to be dealt with without destroying the good commercial relationship between the parties. This professional attitude increases the possibility and the ability of the parties to settle the dispute internally.
- The tendency to push the dispute to the limit, as mentioned above, will sometimes have the consequence that a court case or arbitration becomes the end result despite the parties' policy to settle by direct negotiations.

Question 5. How do you see the trend in frequency of litigation as the final solution?

In their replies the Lawyers mention the following trends:

- The frequency of litigation might be slightly increasing after a number of years where the trend was decreasing. Litigation seems to be preferred over arbitration in a number of jurisdictions, as litigation is seen as less slow than before – and much cheaper than arbitration.
- The trend is, that there is an increased interest to see what would be the outcome of litigation, irrespective of the facts that the costs involved in big court cases and arbitrations are significant. The cost of litigation and arbitration has become a more ordinary cost factor in the business than before.
- Contracting parties do not, as a principle, like the idea of solving a dispute by litigation. However, as outlined above under question 1, there is a tendency to push the dispute to the limit and at least to start a court proceeding in order to achieve the most favorable result. When realizing that such a court proceeding is a very cost- and time-consuming battle, most parties change their attitude and aim at a settlement. The attitude of the contracting parties is very often different, if there is a long lasting business relationship between them and if there is a lot of experience in cooperating with each other.

Question 6. Is conflict prevention by better and more careful contract drafting, involving both parties one of the ways to prevent contractual conflicts?

All the replies to this question are affirmative and some stress the importance of improved contract drafting

- Contract negotiations should create absolute certainty as to the parties' position in a number of vital areas. Of course there is no such thing as a contract answering all questions which might potentially come up, but both parties having a clear understanding on what is agreed (as described in the contract) is certainly always an advantage.
- Disputes arise where the contracts are poorly drafted leaving too many incidents up to the parties' own interpretation.
- A poorly drafted contract leaves room for misinterpretation and misjudgement of the applicable rules. It is, therefore, very recommendable for both parties to make a strong effort in drafting a clear and precise contract and to involve legal advisers in this work. By this, potential risks and conflicts are addressed and can be dealt with already during the negotiation process. It is a far better investment to spend more money in this phase of a contract than to spend at least the same amount of money in a conflict arising because of a poorly drafted contract.

Question 7. Conflict resolution by mediation (involvement of a third party going in between) before litigation – what is your experience with this method?

In their replies to this question the lawyers mention the following trends:

- Early focus on the conflict and the ability of lawyers to identify, where exactly the parties are incapable of meeting without help, is of extreme value so that help is only sought on the matters where the parties are completely locked into their separate positions.
- There is definitely room for an alternative dispute resolution such as mediation and in some jurisdictions and also under some standard contracts it is a precondition for going into arbitration that the parties have gone through an alternative dispute resolution process. The success rate is expected to be approximately 50%.
- International contract law provides for several possibilities that can be included into a contract in order to solve conflicts within a reasonably short period of time, still during the project phase such as Dispute Adjudication Boards, Pre-Arbitral Referee proceedings or a so-called Arbitrator's Expert Opinion Procedure. Mediation is another mechanism for short term solutions next to the aforementioned ones. As per today it is still difficult to persuade the parties to include mechanisms for an alternative dispute resolution into the contract. This might change, however, in the near future. We are seeing more and more consulting companies such as CEDR (Centre for effective dispute resolution) and companies with great technical expertise such as Germanischer Lloyd that offer project-accompanying dispute resolution methods, e.g. with regard to offshore-projects.
- Author comment: Mediation is considered an interesting method of conflict resolution but apparently used very little! Why?

Question 8. Conflict resolution by mediation (involvement of a third party going in between) during and parallel to litigation – what is your experience with this method?

In their replies to this question the lawyers mention the following trends:

- Apparently this happens unfortunately often too late, – once the Parties have come as far as preparing a writ, present their witnesses etc. they seem less inclined to accept an offer for anything but a mock mitigation process.
- There is no difference on whether the alternative dispute resolution is sought before or parallel to litigation. Parties with a professional attitude can distinguish between the two tracks for obtaining a solution to the dispute.
- It can be observed, that in Germany we find more and more judges that are also qualified as mediators and that in very difficult, extremely unclear and complicated cases German courts encourage the parties to carry out a mediation process together with one of these specifically trained judges.
- In an arbitration proceeding it is often easier to persuade the parties of the advantages of a settlement than in a court proceeding. The reason for this might be that arbitrators are usually very experienced judges or lawyers with a professional view also on the commercial aspects of a dispute and a project. High quality arbitrators are usually, more than judges of state courts, trying to find a settlement that is acceptable for both parties. Also, in general they are more outspoken in terms of the prospects for both parties in an arbitration.

Question 9. Regional differences in above mentioned trends – what are the characteristics?

In their replies to this question the Lawyers mention the following trends:

- Asia: There seems to be much more reluctance to settle disputes in court, – gentlemen are here understood to be able to settle these issues out of court. Mediation is still quite uncommon!
- Europe – West: Mitigation has become increasingly more popular, – it still seems to be the general trend that litigation, arbitration and mitigation is evaluated on its commercial merits; what ever seems to bring the wanted results in the most expedient and most reliable manner will be chosen, – with due regard of course also for the costs involved. The use of mediation to settle a dispute is quite common in the UK. Mediation is less used on the continent. In Western-Europe mediation can still be considered as a quite unusual method. It is in particular uncommon to integrate rules for a mediation process into the contract at least as far as the continent is concerned.
- In the American countries alternative dispute resolution is sometimes mandated by law e.g. Canada. The system for mediation is very well developed in North America. Companies and lawyers in the U.S.A. seem to be open-minded with respect to alternative dispute resolution methods. It is not uncommon in the United States to provide rules for a mediation process in project contracts.

3.8.3 Asian Contract Drafting as Observed by Europeans

Project business in Asia seems sometimes to be very USA or European style and sometimes to be very Asian style and following local traditions. An example was experienced by one of the Authors when visiting "Japanese engineering – contracting houses" looking for technology transfer partners. During the same week the Author both witnessed a very traditional Japanese style reception with Komatzu (long and very polite traditional opening talk of everything except business) and a very American style (straight to the key point) with Japan Gasoline Corporation.

The traditional Asian business system is apparently based on a number of principles of which the more important are:

- The non confrontational method – the polite, smooth and friendly mode of operation
- The hierarchical "clan" and family system
- The wide power distance between the mighty "clan" leaders and their subordinates

Asia appears full of surprises at least at the outset. At first glance it seems to be very American or European but in reality it is traditional Asian. In Europe the engineers right away tell their managers all the so called "truths" they come across in their daily work, whereas in Asia they do not tell their managers anything problematic before it is too late to do much about it. Or they have learned the European way and are telling their managers what these want to hear and that everything is fine. "No problem – sir!"

The traditional Asian contract drafting and negotiation method might be characterised by the following factors:

- The contractor has to make his quotation work and bidding based on very little tender information or rules, with virtually no bidding procedures nor evaluation system published beforehand
- Low price is the overwhelming objective and competitive parameter
- Contract conditions are not published or discussed prior to selection of 2–3 bidders for negotiations. The Owner's unwritten conditions might already be known by the experienced bidders
- Focus is primarily on price, on the scope of work and on the time schedule
- When price, time schedule and then scope of work in principle have been agreed or at least fixed for the time being – meaning that the contractor can not go back on them – then the Owner typically presents his contract draft for acceptance by the contractor with very few and minor changes allowed!
- Some of these contract drafts are poor or biased in favour of the Owner
- Consequently the rational two-sided contract drafting work and negotiations aiming at producing a common manual for the project implementation has poor conditions but it is not impossible for the courageous bidder with a competitive price to change and improve the conditions somewhat
- Despite the above negative contract drafting factors the project implementation normally goes relatively smooth because of the magnificent flexibility of the Asians during project execution
- In most cases the project gets finished more or less on time but sometimes with loss of profit or negative profit for the "poor contractors". It is expected to be regained in future projects
- In other cases the contractor makes his budget margin by an execution without problems and due to improved procurement prices

- The risk is of course, that fundamental adverse material or financial changes of conditions (other than contract conditions) appear and then the contracting company's future existence might be at stake!

3.8.4 American Contract Drafting as Observed by Europeans

North America seems to be the lawyers' territory and litigation is a common consequence of conflicts. But in reality many project contracts are executed and conflicts solved without litigation because it is a cumbersome and expensive way.

In North America contract drafting has a high priority and is a very serious matter with lawyers often involved in the drafting process. This seems generally to lead to more professionally worked out project contracts. This has the consequence that the relation building between the parties and development of the contract as an implementation manual is more in focus.

The often used system of paying a part of key staff's salary as a performance related bonus is in the Authors' opinion a dangerous factor for the endeavour to build relations and a practical manual type of contract for execution, because such an activity might be considered waste of time and effort.

Successful project work requires genuine teamwork between different disciplines and management levels especially in the quotation and contract drafting stages, and bonus systems for key players in contracting can be very dangerous as it focuses on the short term personal gains instead of the long term company perspective.

South America seems to be in their contracting practice much more like South European "old style". The major Owners have developed their own standard contract conditions, which can only be modified to some extent, if the contractor has very good reasons. The old type of standard conditions like UN ECC 188 is also commonly used by the smaller Owners.

Generally contract drafting is a sensible area and very much related to confidence or lack of confidence. In South America commercial confidence is built up by long term (min. 5–10 years) personal relations and by the contractor's/supplier's local long term presence close to the Owner. Some business people of South America seem to possess a "down under syndrome" as some Australians' do.

3.8.5 European Contract Drafting as Observed by Europeans

In Europe there still is a French contracting tradition, an English tradition, a German tradition (VOB) and a Scandinavian tradition (NLM) each with its own characteristics and standard conditions. FIDIC-type of conditions are used in certain sectors in Continental Europe but little known in UK and North America.

But although the European Union has harmonized the contract law in general, little has been achieved with respect to project contracts and standards.

Whereas local contracting within a country is normally based on a few sets of standard conditions, the international contracting in Europe is very dispersed with respect to using the various contract drafting methods:

- Based on standard conditions as FIDIC or ORGALIME
- Individually drafted contracts
- Owner's standard conditions
- Contractor's standard conditions (seldom)

3.8.6 International Standards and Good Practises

Based on the observations of regional differences in contract drafting, it is the Authors' opinion, that the main issues in international project contract drafting are the following:

- Does the contract drafting take its outset in international standards and good practises or in national contract law and related legal practises?
- Will the contract drafting focus on obtaining a well functioning manual for execution or on preparing a legal weapon in case of litigation?
- How should the contract drafting be organised in order to achieve a workable contract which avoids conflicts?
- How can the contract drafting activity be separated from the contractor selection and negotiating process where both parties are focussed on prices, time and scope of work? The contract drafting activity has in our opinion to focus on minimizing frictions between the parties during project implementation, thereby preventing conflicts.

The most important examples of international standards useful for international project contract drafting are:

- INCOTERMS
- D/C Standard or Uniform rules
- ICC Arbitration clause or similar
- ICC Pre-arbitration Expertise clause or similar
- Standard general conditions as FIDIC or ORGALIME
- CE Standard, DIN Standard, British Standard, American Engineering Association Standards, etc.

The best examples of good practices in international project contracts are:

- Supplier/contractor has the obligation to define his scope of work clearly to protect himself
- Owner/employer has the obligation to define the plant performance in the way he wants it – supplier/contractor then needs to protect his performance guarantee by defining the input to the process in order to be able to fulfil the performance guarantee
- Supplier/contractor has the obligation to perform the project implementation on specifications and on time irrespective of obstacles except if they are significant, notified and documented hindrances – anyway mitigation of losses and effort to catch-up on delays are contractor obligations

- Supplier/contractor has the obligation to perform instructed/agreed variation work or work in dispute (variation order work versus work inside the Contract scope) within the agreed time schedule
- Scope of work is defined both by general functional requirements (fit for the purpose) and by specifications and both have to be fulfilled

3.9 Conclusion

From a project conflict prevention or resolution point of view the preparation and drafting of the contract are crucial. The better and more thorough preparations are made the less risk of later conflicts there are. We will now summarize the most common contract drafting and preparation methods.

An individually made Contract, which is drafted by the parties for the specific project, is in our opinion the best method. It should be used when the parties realize that the Tender Documents as well as the successful contractor's bid need modifications and instead of adding these modifications, they decide to write a new contract (including a number of annexes) reflecting the agreement of both parties to work together under the agreed conditions. The purpose and intended consequences are that both the Tender Documents and the Bid are replaced by the Contract.

This drafting activity might take from a few days to a few weeks of extra work by both parties, which actually are next to nothing, compared to the risks involved. Furthermore the common drafting effort will enhance cooperation during project implementation.

The other very commonly used method is "the sophisticated order acknowledgement type" or "short contract agreement type" with full use of Tender Documents and the Bid. In this case the full contract will consist of the following individual documents:

1. Agreement signed by the parties, specifying the order of priority of contract documents etc.
2. General Conditions
3. Special Conditions
4. Minutes of Meeting between parties regarding clarifications and changes to Tender Documents
5. Tender Documents
6. Successful Bid
8. Other documents regarding taxes, shipping, custom clearance, standards etc.

The advantage of this method is that it is fast and simple with respect to avoiding lengthy discussions before signing of the contract. But the disadvantage is that individual documents are prepared by different parties at different times for different purposes and therefore they are very difficult to combine and this leaves a lot open to interpretation. Furthermore the definitions of important notions can not be made or be made only with great difficulty. In most cases these important contract definitions are not included and this increases the conflict risk because they are interpreted differently by each party.

A number of excellent Standard Contract Conditions of a much higher contractual quality than most of the individually drafted contracts exist. They might be a solution to the dilemma mentioned above. They are more or less neutral to the conflicting interests of the parties and internationally well known and accepted.

The best known examples are a series of Standard Contract Conditions prepared and published by the Federation Internationale des Ingénieurs Conseils (FIDIC) e.g. Conditions of Contract for Design, Build and Turnkey. In our opinion FIDIC Standard Contract Conditions are very comprehensive but impractical because the system and the wording is far too general and sometimes very complicated for international project staff.

Another example is ORGALIME SE 2000 General Conditions for Supply and Erection of Mechanical, Electrical and Electronic Products. ORGALIME is the European Engineering Industries' Organisation which might disqualify the standard as being biased. It also contains the same disadvantages as FIDIC although to a much lesser degree.

The third example we want to mention is ENAA (Engineering Advancement Association of Japan) "Model Form International Contract. Alternative Form for Industrial Plants". This is from an international project contracting point of view the best standard seen for a long time. It might be considered biased because it has been drafted by contractors. But it can be used and changed as a draft and it is relatively simple.

In our opinion the most important disadvantage of these Standard Contract Conditions is, that they tend not to be studied and understood by the project staff to the same degree as the contracts drafted individually.

In any case, the general conclusion is, that much more management and specialist attention, effort, resources and time are needed in order to obtain better and more workable project contracts in order to reduce the frequency and consequences of disagreements and conflicts.

On specific areas progress has been achieved by the use of international standards such as INCOTERMS. Further development of such specialised standards is a way forward to improve the project contracts.

3.10 Questions on Chapter 3

1. Name the documents of a project contract and explain in a few words their interrelation and function.
2. Name the main issues in a project contract and explain in a few words, how an agreement can be made and documented in writing.
3. What are the 5 most important responsibilities of the main contractor in a project turn-key contract?

4. What are the 5 most important responsibilities of the client/owner/employer in a project turn-key contract?
5. What are the 5 most important responsibilities of the main process equipment supplier in a supply + supervision contract?
6. What are the 5 most important responsibilities of the client/owner/employer in a supply + supervision contract?
7. Describe the similarities and differences between the project turn-key contract and the main process equipment supply + supervision contract.
8. What is the point of risk transfer in the project turn-key contract and in the main process equipment supply + supervision contract?
9. Mention the 5 most important issues that can cause a conflict in a project turn-key contract and in a process equipment supply + supervision contract (i.e. 2 times 5 issues).
10. Mention the 5 most important contract drafting measures that can be used to prevent a conflict in a project turn-key contract and a process equipment supply + supervision contract (i.e. 2 times 5 issues).

References

www.iccwbo.org International Chamber of Commerce.
www.fidic.org International Federation of Consulting Engineers.
www.orgalime.org European Engineering Industries Organisation.
www.enaa.or.jp Engineering Advancement Association of Japan.

4 Preventing Conflicts by Application of Psychology

Written by Psychologist Rikke Rye Hahn, Svendborg, Denmark

Abstract. Why are we including chapters about psychology in a book that deals with conflicts in projects? The answer is easy: behind all conflicts there is also a facet of psychological issues. In other words: Without psychological dimensions, the conflicts and disagreements would be much less pronounced.

In this chapter we will show, that personality, basic beliefs and values on the individual and organizational level play an essential role in leading to differences, misunderstandings – and finally: conflicts. Furthermore it will be shown that conflicts can not be prevented or resolved without taking the psychological aspects into consideration. We will present a model for developing "conflict competency"; knowing how conflicts develop and escalate, how to prevent the conflict from building up, and how to resolve a conflict, which has escalated. Using these tools a project manager, will be well suited to deal with conflicts before, during and after they have occurred. This chapter describes how psychologically to deal with different conflicts as illustrated by a real case story using mutual understanding, the psychological contract and the psychological check list presented.

Key words: Application of Psychology, psychological dimension, personality, conflict competency, conflict escalation, conflict dimensions, dialogue, mutual understanding, personality traits, communication, partnership, conflict competent leader, mediation, alternative dispute resolution, ADR, dynamics of conflicts, reactions to conflicts, constructive conflict communication, destructive conflict communication

4.1 Case Study: The Fertilizer Plant

In order to show the impact of psychology on the development of a conflict, we shall first perform a case study, which will help us understand the motives of the acting people.

A) Description of the Conflict

- A fast growing European main contractor sold a fertilizer plant consisting of a number of process units to a Middle East owner. The plant was based on process know how and equipment specifications and prices from other European process suppliers and contractors. The main contract was an adapted FIDIC type turn-key contract with the Owner's Managing Director as "The Engineer" holding strong approval and change powers of the fit for purpose type.

- One of the units was foreseen to be supplied by a European engineering house. During the prescribed basic design approval procedure, which took place after effective subcontract (closely linked to the main contract by back to back provisions) The Engineer/Owner succeeded in increasing the design parameters, resulting in larger and more expensive main process equipment. The consequences were cost increases of about 20% and extra execution time.
- One subcontractor of the European Engineering house did not accept the increase of capacity of the unit to be provided by him and asked for extra payment. The European Engineering house (Main Contractor) did not accept.

B) The Parties' Approach to Resolution of the Conflict

- The case was handled by ambitious middle aged engineering directors both on Main Contractor's (Mr. A) and Subcontractor's (Mr. B) side and an agreement of who would bear the extra cost for the increase in capacity was not found. In view of the missing agreement the frustrated Subcontractor threatened to stop work, till his extra cost would be covered by the Main Contractor. Finally negotiations started, but they were carried out in an unfriendly, inflexible and quite formal way with lawyers participating as from the beginning in the negotiating teams on both sides. No solution was obtained, and more than 2 years after execution of the contract had started the subcontractor decided to start an ICC arbitration.
- Three years later the unit was completed by the subcontractor with a substantial delay and accepted by the client and the main contractor. During execution numerous conflicts and arguments came up, of which only very few were resolved. The arbitrators decided the case by only accepting a minor price increase in favor of the subcontractor, but placed the main responsibility for the delay on the subcontractor, who had to pay substantial delay costs, which were higher than the contractual delay penalties. The total conflict costs for the subcontractor were significantly higher than the originally claimed amount. The main contractor had to bear the slight price increase for the larger equipment and part of the cost for the legal proceedings.

These are the external facts of the case. As we see, the negotiations were carried out from the start in an unfriendly, inflexible and formalistic way, with lawyers on both sides. Thus, what was meant to be "negotiations", sounds more like "wars" – both parties fighting for their own rights, and with no interest for the other party's interests and rights.

How could the above mentioned situation have been avoided? Could the situation have been handled in a more effective way?, without people getting hurt?, losses on the financial and professional as well as on personal and emotional levels being avoided?

When introducing psychological methods to prevent and resolve conflicts, we take the outset in the above illustrated case. This will lead to a different advice on what to do in order to prevent and resolve conflicts in projects. Some of this advice are centered around creating individual "conflict competency". In order to develop these skills, a person – or even better: the organization – must have:

1. Understanding of the dynamics of conflicts
2. Understanding of one's own and the other party's reactions to conflicts and thereby gaining knowledge about the persons involved in the project
3. Fostering of constructive responses to the conflict and reducing destructive responses
4. Understanding of organizational differences - knowledge about the different organizations intervening in the project
5. Knowledge about how to work out a "Psychological Contract", in order to make both parties cooperate in preventing the conflict
6. Knowledge about different conflict resolution strategies; i.e. mediation or litigation, mentioned in Chap. 6.

These six aspects will be the headlines of the chapter sections – including ideas and specific suggestions to prevent conflicts during projects.

4.2 Understanding the Dynamics of Conflicts

In this section we will impart the understanding of and the interest in conflicts in general, thereby the reader should become able to identify a potential conflict and do something about it at an early stage.

The Austrian conflict expert Friedrich Glasl (Glasl 1999) has made a theory saying that all conflicts which escalate, will go through some general steps, which will be discussed in the following sections. Each step has its own dynamics.

At the first level both conflict parties can still win (Win Win). At the second level a party loses, while the other one wins (Win Lost) and in the third level both parties lose (Lost Lost).

Interestingly enough one can analyze the most different conflicts by using this model: From wars and conflicts between states to conflicts between colleagues or pupils. So this also goes for analyzing conflicts in projects.

4.2.1 Level One (Win Win)

Stage 1: Tension

Conflicts begin with occasional collision of opinions leading to tensions. It is an everyday event and not noticed as beginning of a conflict. If the conflict develops nevertheless, the opinions become more fundamental. The conflict could have deeper causes.

In the considered case study of Sect. 4.1 stage one takes its outset in the two parties not being very fond of each other, and not being certain about the other one's intentions. Unfortunately, they do not recognize this. Instead, they just go on as usual; working on their own interests – but at the same time a fundamental suspicion towards the other person develops.

Stage 2: Debate

Thus, when a concrete conflict occurs – the question of who should pay the extra costs of the equipment and for the loss of time – both Mr. A and Mr. B hold on to their basic mistrust towards each other, because the solution can affect their respective careers. This makes a resolution less likely to happen. None of them are willing to compromise; just pointing fingers at each other "it's your fault!" Compromising is for each of them seen as admitting to be part of the mistake; and would create a feeling in each of them of having "lost".

Friedrich Glasl (Glasl 1999) explains stage 2 as: Each partner, thinking to be in his own rights, tries to convince the other of his own arguments. Diversities of opinion lead to a controversy. One tries to pressurize the other.

Stage 3: Acts Instead of Words

Each conflict partner increases the pressure on the opposing party, in order to become generally accepted. Discussions are e.g. broken off. No more communication takes place and the conflict intensifies.

In the case study of Sect. 4.1 this behaviour is illustrated by Mr. A and Mr. B communicating less with each other, and more with other people – internally as well as externally – ; about each other! This "talking badly about the other" just moves them even further away from a mediation meeting – which could still have prevented the conflict from developing even further.

4.2.2 Level Two (Win Lost)

Stage 4: Coalitions

The conflict is intensified by the fact that one party looks for sympathizers with his position. Since he believes in his own rights, he can "denounce" the opponent. The arguments no longer concentrate specificly on the subjects of the dispute, but the parties introduce adjacing subjects, thus trying to win the conflict. The major interest now is, that the opponent loses.

As described, both for Mr. A and his company, and Mr. B and his company, it became even more important at this stage to prevent the other party from winning, than to create an atmosphere, where both parties could win. This initiated the unfriendly and formalistic approach , as well as the communication through lawyers only, in stead of meeting face to face. The conflict escalated – and as Friedrich Glasl (Glasl 1999) describes it: at this stage it is no longer possible to solve it without the intervention of an objective third party; a mediator.

Stage 5: Face Loss

The opponent is to be destroyed in his identity by all possible means. Here the confidence loss is complete. Face loss means in this sense loss of the moral reliability.

A lot is at stake here for Mr. A and Mr. B. They risk loss of face in their own companies – and with the opponent. The more this threat seems realistic to them – the further they will remove themselves from communicating, and resolving the conflict.

What is surprising here, is that the companies behind Mr. A and Mr. B let all this happen. Still, such a situation is not so rare: the company behind a person can actually strengthen the conflict, because of internal high esteem for the own employee. Both Mr. A and B were, as mentioned, very ambitious, and tried hard to live up to the expectations of their companies. Thus, loss of face seemed a very big threat to both of them, this made a resolution even more difficult to reach.

Stage 6: Threatening Strategies

With threatening each other the conflict parties try to control the situation. It is to illustrate their own power. One threatens e.g. with a demand (10 million Euros), underlining the request with a heavy sanction ("otherwise we leave the construction site"), one intensifies and one supports the request with all the sanction potential (explosive show). Here the proportions decide on the reliability of the threat.

4.2.3 Level Three (Lost Lost)

Stage 7: Limited Destruction

Now the opponents try to sensitively harm each other and the opponent is no longer taken as a human being. Starting from here one's own damage is now disregarded, and the only objective is to inflict damage upon the opponent, the greater the better.

Stage 8: Splintering

The opponent is to be destroyed with destructive actions.

In the case study of Sect. 4.1 the threat and destruction became very realistic; to stop the project was no longer a threat, the worst case happened and the project was stopped. At this stage, none of the opponents was able to stop or control the escalation.

Stage 9: Together into the Abyss

One takes into account their own destruction, continues the attack on the opponent, in order to defeat him.

4.2.4 Evaluation of the Friedrich Glasl Model with Respect to Conflict Escalation

The model describes how two parties in conflict behave. Solutions to the de-escalation are not offered in this model. In particular, conflicts, of which both conflict parties think to take advantages, do not appear possible to be de-escalated (e.g. aggressive acts on the territory of a State by another State, paternal handling of a common child after a separation, withdrawal from citizenship by a Country, mass dismissals of a company with the aim to improve its shareholder value), when one or both parties deliberately use escalation as their strategy.

The capability of a worth-free recognition and elimination of forces driving the conflict should belong to the standard repertoire of high-level managers, advisors and social workers.

We shall now look at how the model of conflict escalation can be applied to the case study of Sect. 4.1.

First, the model of conflict escalation can be used in different ways. What is needed is first of all to create a common language about conflicts within an organization and with outside parties. If a colleague or external business associate refers to a conflict as a stage 4-conflict, everyone knows what is meant.

Second, the model of conflict escalation offers a unique possibility to separate the persons from the problem. When, for instance, a stage 4-conflict is under way (as illustrated in the case study), the two parties directly involved (Mr. A and Mr. B) are not alone the problem. The problem is, that the conflict is existing, and that it has developed thus far. By using the model of conflict escalation, it is no longer necessary to spend time on searching for responsibilities, but it is required to look at the future rather than the past. The focus has to be on, how we cope with the future instead of blaming our opponents, for what has happened in the past.

Third, the model of conflict escalation gives us an opportunity for making a thorough conflict analysis: Knowing, how far you or someone else is involved in a conflict, can help you identify what to do at the actual stage of the conflict. What strategy should be used? These strategies will be further described in Chaps. 5 and 6.

4.3 Dimensions of Conflicts on Project Level

When wanting to understand the dynamics of a conflict, you may ask yourself: What is inside a conflict, what does it consist of? In relation to the case study of Sect. 4.1, you may wonder why it escalated that far. Understanding the dimensions of the conflict may give us some answers to this question. Whether we ourselves are entangled in a discord or we stand outside it and try to help, the problem often seems endlessly confusing. There are so many elements, so many feelings, where to begin? To help establish a rough overview we list four basic dimensions. These are:

1. *Instrumental dimensions*
2. *Dimensions of interest*
3. *Dimensions of value*
4. *Personal dimensions*

The instrumental dimensions are at work when we disagree about objectives and methods: what to do and how to do it. At this level negative sentiments and reproaches may not yet pollute the relation. We just disagree and have to find a solution, in order to get on with the matter. We must solve the problem. We have this type of conflict very often; they seldom lead to animosity and are often useful for creative decision making.

Dimensions of interest occur, when there is a competition for resources, which are sparse or appear to be sparse. The resources could be money (e.g. the price of something), time (e.g. spare time), and space (e.g. who's allowed to take up the most space?). At home, it may be the allocation of rooms, housework and leisure time. At work, the dispute is often about working plans, facilities and wages. At a larger scale, there is the fight for power, territories and economic dominance. Globally there is a struggle for basic necessities of life, water supply, and other resources of nature. In conflicts of interest we have something solid between us – and therefore we may negotiate and find solutions.

Dimensions of value appear in a conflict, when values, which are precious to us are at stake. These are values, we are willing to stand up for. Such values might be morally antagonistic. What is right, and what is wrong?, they might be values of our traditions, of our religion, our political beliefs and our dedication to human rights.

The personal dimension of conflicts often infects our personal existence and everyday life and creates vast confusion and suffering. Here, deep and sometimes hidden feelings play the leading role, and the parties become uncertain and vulnerable: do the others regard me as somebody? Does anybody at all see me? Can I trust them? Are we kept out? Do they despise us? Are we being respected? Some of these feelings have to do with past experiences and forgotten happenings of our childhood.

The fusion. In real life these four dimensions are often completely entangled. In the case study described in Sect. 4.1, we saw two persons; Mr. A and Mr. B, each representing a company, having a dispute that should cover for common losses. This may look like an instrumental conflict. But it may, at the same time, be a conflict of interest and even deeper – in fighting for power or esteem. Thus, also value dimensions and personal dimensions are involved.

To a lot of people, it may seem a little ridiculous that the two companies, represented by Mr. A and Mr. B didn't just solve the conflict. Why risk a loss bigger than the original claimed amount, in order to "win" a conflict? The essence of the problem, and the reason why the conflict wasn't solved, is probably that they were dealing with a conflict which took place at all 4 dimensions as described above – but they were trying to solve it, only focusing on the first two dimensions; dimensions of interests and dimensions of values.

This is a common mistake when trying to deal with conflicts: people negotiate, instead of communicate! People focus on the visible conflict – but ignore the very important value dimensions and personal dimensions. This will only result in a superficial solution – but not in a deeper and lasting resolution.

In the eighties, when the USA and the USSR were negotiating disarmament, the two governments were not able to agree upon where to meet or upon the shape of the negotiation table. This may look like an instrumental conflict, but it was probably more a conflict of interests like economy and world supremacy. Maybe even a personal conflict – statesmen also have strong feelings although they claim to be 'objective'.

But if it is true that the dimensions in fact always are merged into one another, what is the purpose of distinguishing? What's the point of a model like this? Why spend time on analysis?

Because in any conflict there will be one of the dimensions that is more significant than the others are. We call it *the centre of gravity*. It is useful to sort out leads from this basis. If there are deep emotional problems in one party or both, then these emotions have to be addressed in a dialogue, and only after such dialogue we can expect the persons to act sensibly and stick to the subject matter. Furthermore, if there are real

Schedule 4.1. The Four Dimensions of a Conflict:

1 – Instrumental dimension
About: Tangible issues like methods, procedures and structures.
Approach: Problem solving.
Desired aim: Solution.

2 – Dimension of interest
About: Allocation of resources like time, money, labour and space.
Approach: Negotiation.
Desired aim: Agreement.

3 – Value dimensions
About: Political, religious, moral values.
Approach: Dialogue.
Desired aim: Mutual understanding.

4 – Personal dimensions
About: Identity, self esteem, loyalty, rejection, etc.
Approach: Dialogue.
Desired aim: Mutual understanding.

and serious conflicts of interest between two parties, then these interests must be addressed, and the discord can not be managed as an emotional problem.

We shall now discuss how to deal with different conflicts. The main issue is, that we have to find out, which is the central and dominating disagreement of the conflict, in order to be able to cope with it.

As shown in Schedule 4.1, instrumental and interest conflicts can be solved by negotiation. We can discuss how to do things (instrumental dimension), and we can discuss what benefits can be obtained with a solution (interest dimension). We can even reach agreements and solutions. But we cannot negotiate our beliefs (value dimension) in order to find a compromise. Nor can we negotiate our feelings (personal dimension). What we can do, is to have an open communication and a dialogue about them. If this dialogue takes place, the result may be, that we come up with a better understanding between the other person and our self. This creates a friendlier atmosphere, which in turn eases the negotiation over the opposing interests and instrumental points of view.

In relation to the case study of Sect. 4.1 we now understand, that the basic fault was, that Mr. A and Mr. B tried to solve the conflict by making lawyers run the negotiation. This could have solved the conflict, had it only been an instrumental one, or a question of interest. This was however not the case. The case was that apparently Mr. A and Mr. B – and the companies they represented – both had some more at stake: professional (and personal) pride, a tradition for winning cases, a longing for proving themselves to be right, and so on. As described in the model above, the aim has to be mutual understanding, when wanting to reach a conflict resolution in a conflict, where value dimensions and personal dimensions are at stake. Understanding the interest of the other person, or of the other company is essential here. And a mutual understanding like this can be reached only through dialogue. No other method will work.

To sum up, a first important step in dealing with conflicts is to understand their dynamics for which we have presented two tools: 1. the model of conflict escalation, and 2. the model of the four dimensions in conflicts. These two models help us to analyse and understand a conflict, and thereupon how to deal with it.

4.4 Understanding One's Own and the Other Party's Reactions to Conflicts

We learned in the last paragraph, that it is mutual understanding, which helps to find a resolution in a conflict involving value dimensions and personal dimensions. Understanding the interest of one self, one's own company, the other person's interests, or the other company's interests is essential here. This can hopefully lead to both parties opening up for negotiation, as described in Chap. 5.

As mentioned in the case study of Sect. 4.1, there is a very important psychological level underneath the action-level: This deeper, emotional level includes all personal, organizational and cross-cultural intentions, needs, interests, and so on. These personal dimensions are invisible – and we can only get an understanding of them – and thus reach mutual understanding – when we ask, what goes on here? "Responding mindfully" to conflicts means: taking the other person's perspective into consideration – rather than holding on to your own perspective and point of view. Typically the parties of a conflict become more and more dualistic; "black and white" in their understanding of the dispute. The issue "Who is right and who is wrong?" seems of greater interest, than actually reaching a mutual understanding of the situation.

Fig. 4.1. Project Manager Seminar on "Understanding the psychology of conflicts"

As we saw in the case study of Sect. 4.1, instead of opening up to each other, both parties of the conflict were occupied with trying to "win" their own case, which included making the other part loose. When trying to win, a person will typically concentrate on his own perspective, preparing an attack strategy rather than concentrating on, what the motives are of the other party. In the conflict of the case study of Sect. 4.1 it would have been beneficial for both parties to reach an understanding of the following:

1) The persons involved (especially Mr. A and Mr. B, as they were the ones to run the project and the negotiations in the first run).
2) The companies involved and their different cultures, interests, values etc. This becomes especially crucial, when cooperating with external parties. This will be described later.

4.4.1 The Persons Involved

It is especially crucial for a project manager – who wants to become a conflict competent leader – to understand his own strengths and development opportunities with respect to a conflict. Better self-awareness enables a project manager to identify where and how conflicts arise due to him, how he might cause conflicts, and how he responds emotionally and behaviorally in a conflict. This understanding will make it possible for the project manager to devise the most effective behavior before, during and after a conflict.

Everyone is aware of characteristics in which he differs from other people. These could be different thoughts about strategic directions or tactics, perceptions about resource allocations, opinions on organizational structure and individual responsibility. How do these differences arise?

Some may be due to a different level of information about a problem. Some can stem from naturally competing interests as departments competing for resources or individuals competing for promotions, or organizations competing for economical gain, as in the case study. In many cases, though, the differences arise from people's different personalities, education, preferences, and styles. Too often these lead to conflict and misunderstandings. The conflict competent leader understands his or her own styles and preferences, recognizes potential strengths and weaknesses related to them, appreciates that other people have different ones, and works to make these differences a source of opportunity rather than a conflict.

Having a broad understanding of differences between people is a tool on how to respond mindfully towards different kinds of personality. It's not the objective here to bring a definite theoretical explanation of personality. Instead we'll enhance some practical tools to recognize different personality traits, in order to be able to match these in a professional (or personal) communication.

4.4.2 Different Types of Personalities

Throughout times many different theories about personality have been made, including categorization of personality. Tools have been made, in order to assess personality, so that we are certain about, which category we belong to. In this book, we refer to an understanding of personality and categorizing of personality, which belongs to trait theories, and "The big Five". For further information, read for instance: (Catell 1965), (Allport 1961), (Hogan 1991). One way of categorizing people, is by using the so called "Personality Indicators"; such as 16 PF-R, MBTI, P.I. and DISC, which are discussed in the literature, e.g. (Catell 1965), (Allport 1961) etc.:

- Level of dominance – drive to exert one's influence on people and events
- Level of extroversion – drive for social interaction with other people
- Level of urgency – intensity of a person's tension and drive
- Level of detail orientation – drive to conform to formal rules and structure

A person who has a high level of dominance we found in the case study in Mr. A. He was described as a very individualistic person who preferred to be in charge, and make all decisions. He was deeply provoked by Mr. B by not doing as he was "told to do" by Mr. A. He is a person who communicates very directly and without paying much attention to people's feelings. Interestingly enough, when holding meetings, he was able to mention all the "right" methods to use: "We must adapt to each other", "We all have to compromise", etc., he just didn't follow these intentions himself. This was probably due to his high strive for "wining" and competing. At the same time, the very fact that he said one thing, but acted in a totally opposed way, was what made it very difficult for Mr. B and other external parties to cooperate with Mr. A.

We can recognize a person, who has a high level of dominance, by his behaviour, which is:

- Independent
- Assertive
- Self-confident
- Self-starter
- Challenging
- Venturesome
- Individualistic
- Competitive

A dominating person has certain needs, and if these needs are not fulfilled, he/she will feel somewhat frustrated. The needs of a dominating person are: being recognised for own ideas, having the right to act independently, being in control of their own activities and work, understanding the big picture, rather than having to care about details, and having the possibility for competing, with him-/herself and others, and in the best case to win.

On the other side we can recognize a person who has a low level of dominance by the following attributes:

- Agreeable
- Cooperative
- Accepting of company policies
- Accommodating the team
- Comfortable with his situation

- Seeks harmony
- Risk averse

What this person needs in order to be as comfortable as possible, is to be encouraged, and get outer reassurance. This person will seek harmony in the group, and will have a tendency to avoid conflicts. He/she would rather compromise than continuing a dispute. This is a very group oriented person, who would like others to take over and lead the group. He/she doesn't like competition, as it also indicates a less harmonic relationship amongst members of the group.

As the next characteristic we find the level of the extrovert. The very extroverted person tends to behave in manners which are:

- Outgoing
- Optimistic
- Selling
- Delegates authority
- Meets new people easily
- Enthusiastic
- With empathy
- Socially poised
- Talks before he thinks

This person's needs are primarily of social character: being accepted and recognised by other people, having status, being in a position where he or she can "sell" good ideas, and being able to network. Building consensus is also something that brings satisfaction to this person.

Mr. B of our case study in Sect. 4.1 scored very high on this factor, which means that he usually could connect to most people. At the same time, it was important for him to prove himself – earning recognition from the outside world. As the project didn't succeed as well as he wanted it to do, he became more and more frustrated – unable to use his otherwise good communication skills. When he wanted to build consensus, he felt rejected by Mr. A – and thus he withdrew himself.

On the other end of the scale, we find the introverted person. A person who is:

- Serious
- Introspective
- Task oriented
- Matter of fact
- Analytical, Imaginative
- Reflective
- Cautious towards new people, Reserved
- Thinks (long) before he talks

This person feels less comfortable, if he/she doesn't have the possibility for introspection, and privacy, in a quiet environment – or isn't recognised for technical or intellectual achievements.

Introverts, who prefer quieter working environments, might find extroverts, who prefer frequent interaction, to be disruptive. If the differences are too big, the level of disruption may itself cause conflict.

At the high level of intensity we find people who are impatient, action oriented and fast paced. Both Mr. A and Mr B were high on this level, which means that they didn't see the point of talking and discussing; sharing experience or asking the other person's point of view seemed like a waste of time to both of them. Rather, they wanted to just start working and acting right away. Both of them needed freedom to change priorities, variety, and freedom of repetition. Therefore they both had a tendency to become very impatient with negotiations.

Persons of high level of intensity are described as:

- Tense
- Restless
- Highly strung
- Driven
- Impatient with routines
- Intense
- Sense of urgency
- Fast paced

Often, people with a high intensity will conflict with those who are low on the same factor. As low-intensity persons need security, stability, and familiar surroundings, they need especially a lot of time before being able to accommodate to new situations. As persons they show with their actions that they are:

- Patient
- Stable
- Calm
- Deliberate
- Consistent
- Comfortable with the family
- Steady

Thus, the differences between high-intensity and low-intensity persons are huge, and can therefore create conflicts.

Finally, we have a person's level of detail orientation, including the person's drive to conform to formal rules and structures. A person who scores high on this scale, has a need for certainty, understanding exactly what the rules are, being able to work down to the last detail (technical type), needing time to dig, train and study, and being recognised for error free work. This person puts pride into doing the job just right. What can be observed about this person is, that he/she is:

- Diligent
- Attentive to details
- Precise
- Organised
- Self-disciplined
- Cautious
- Conservative
- Conscientious
- Specialistic

What we do not know about Mr. B is, whether he was exactly this kind of person; very precise and quality oriented. Not very surprisingly, Mr. A was actually the same. This trait made them both very proud and righteous when it came to explaining or fighting for their own points of view – and less flexible or open towards the other person's version.

Had one of them been a bit lower in this trait, it might have been easier to negotiate. A person who is very low on detail orientation tends to behave in ways, which are:

- Informal
- Tolerant of risk or uncertainty
- Freely delegates details
- Uninhibited
- Non-conforming
- Casual
- Disorganised
- Undaunted when criticized or rejected

There are no right and wrong preferences or personality traits; moreover, people are not always bound to act in line with their preferences. It is helpful to recognise our own preferences and acknowledge that other people may vary from these. If you clash with someone, stop for a moment to consider whether it may be stemming from these different preferences. If so, you may want to reach out in a manner that acknowledges the other person's preferences and find a way to communicate effectively with him or her.

As a project manager – and as a conflict competent leader – you will hopefully recognise that these differences can be a source of organisational strength. It is helpful for a project to have the flexibility that comes from having some people who work well with details and others who prefer to look for the big-pictures and principles. The key for the competent conflict leader is to recognize and value differences and make sure that others in the project – or external parties – do too. Once the differences are recognized, it becomes easier to overcome the frustrations and use the strengths inherent in having diversity of preferences.

As personality plays a role in our individual way of responding to conflicts, so does our way of acting and communicating, when facing conflicts. The influence of communication and behaviour on conflicts will be discussed in the next paragraph.

4.5 Fostering Constructive Responses to Conflicts

4.5.1 Typical Ways of Acting and Communicating when Facing Conflicts

"As long as human beings have a conscience and intellect to think about the future, definitely there will be conflicts. Conflicts are made by human beings and methods to solve them must be created through human intelligence. It is wise to solve the conflicts through dialogue, not through weapons."

HH. The Dalai Lama, Dharamsala 29.11.2001, (Hammerich 2001)

In his book "Managing Differences", D. Dana (Dana 2001) mentions the "retaliatory Cycle" as the trigger for most conflicts arising. Here, he suggests that most conflicts or arguments have a common anatomy: a triggering event, the perception of threat, defensive anger, acting out, and repetition. A triggering event is any behaviour (may or may not be intended as hostile) which is perceived as threatening or hostile by the other person (the receiver). This perception of threat is accompanied by a natural emotional response to the perceived threat he calls defensive anger. The acting-out phase is critical. It is here that the person feeling threatened behaves in ways that either cause the conflict to escalate through distancing, avoiding, or yielding. Such behaviours will be perceived by the first person as threatening – and the circle has started and will be repeated endlessly.

Certain styles of communication seem to be linked to these destructive behaviours. And others are in it self creating constructive dialogue – leading to solving the problem or conflict. As a competent conflict leader it is necessary to be able to identify these – in order to avoid using them yourself, and in order to recognise a potential conflict when listening to these styles of communication in others.

4.5.2 Destructive and Constructive Conflict Communication

Active destructive communications emerge, when individuals overtly respond to a conflict situation. These ways of communicating require some effort and they almost always escalate the conflict. The active destructive behaviours are among the most toxic responses one can have, and will always threaten to ruin the working relationship.

Constructive conflict communication is, on the contrary, opening, creating dialogue, and fostering an atmosphere of acceptance and respect. In order to lead to constructive communications, we should ask us the following questions:

Winning at all cost or seeking an acceptable compromise?

As we saw in the paragraph describing different personalities, some personality types are very competition oriented, wanting to win all the time. To persons like this, seeking a compromise can be extremely difficult, yet also desirable, as for all persons. When the other party recognises that you are not fighting to win; that you are not trying to eliminate him/her, but that you are trying to work out a plan that can satisfy both his and your own interests, he will ease up, let go off the defences, and help seeking an acceptable compromise. This is the only way of creating, what has popularly been called a "win-win" situation. In these situations nobody looses, as both have been enriched through the process of solving the conflict.

Interrupting or listening?

Why do we interrupt? Probably to maintain and force through our own perception of reality and not let the other person's, who is in our opinion wrong, prevail. But however painful it may be to listen to the other person's version of the story, it is still necessary, if the conflict is to be eased, and how our mind can expand when it happens! A lot of conflicts are based on misunderstanding the universe of the other party. Often we forget to ask the other person what he/she thinks. Maybe we are afraid of negative opinions. Luckily the truth about what he/she thinks about us or the situation is almost always less threatening than in our imagination.

Ignoring or showing interest? The conflict escalates when we treat the other person without respect. It is good for everybody, to be treated with respect and interest. Not just by the spoken language, but by the attitude, body language, and eye contact.

Rhetorical or open questions?

The open questions are inquiring – we already know the answer to the guiding ones. They are part of our verbal combat. Quite often it is not even the question itself, but the tone of voice and hence the attitude that makes the question guiding or open.

Blaming or expressing one's wish?

Blame is widely used as a way of communication, but the outcome is generally doubtful. It is often more useful to express, what you wish or need, than to blame someone for not giving it to you. When blaming, you only make the other person defensive – and create a hostile atmosphere.

Generalising or being specific?

In abstract language we often use words like the general "one" or "ought to". We grant rules and views to the other party and make the responsibility vaguely general. We use the words "always" and "never", instead of referring to the specific case. Generalisation gives the statements an air of supremacy, which provokes the other person's opposition. In the specific language we stick to the actual case.

Past or present?

It can be useful to disentangle the facts of the past. After we have been strongly violated our past differences must be spoken out and recognised, before we can forgive. Also in everyday discords we have to clarify what went wrong in the past. But dwelling

Schedule 4.2. Comparison of Constructive and Destructive Communication

Constructive conflict communication	Destructive conflict communication
• Genuine regret & forgiving	• Superficial excuse
• Trying to learn	• Trying to win
• Expressing own concerns and needs	• Blaming the other
• Showing real interest	• Neglecting the other
• Explaining own views	• Ignoring opposing the other's facts
• Listening to the other's story	• Interrupting the other's story
• Calming & reassuring (Future relations)	• Threatening
• Sticking to facts	• Exaggerating, generalising
• Being sincere	• Using sarcasm
• Expressing one self	• Defending one self
• Listening to the other	• Ignoring the other
• Attacking the problem: What to do?	• Attacking the other: Who to blame?
• Frank language	• Rude or evasive language

in the past and nagging about it may escalate a conflict. Talking more about possible steps to take in the future is more fruitful and less controversial.

Attacking a person or the problem?

The important thing is to distinguish simply between the act and the person acting. This makes it possible to confront the person in a more adequate manner; commenting on the act, rather than on the person: "how you performed the job wasn't good enough… I would like you to do like this in the future…" – is a lot more "acceptable" than: "You are not good enough". We are not able to change, as long as we are being defined negatively – and thus; attacking the person will only make the conflict escalate.

To sum up; what identifies constructive versus destructive conflict communication is summarised in Schedule 4.2.

4.6 Understanding Organizational Differences

Especially when working with other parties, companies are facing risks of conflicts, based on misunderstandings because of cultural differences. This is today common knowledge to scientists within the field. L.N. Marguerre (Marguerre 2000) has developed a tool to prevent and solve conflicts between two or more parties when they want to cooperate on a project. She claims that the biggest challenge for the companies is to ensure good communication relations between the two companies, and to avoid conflicts based on misunderstandings. What creates a good relation is openness, direct communication and honesty. In other words, the companies need as much understanding about their different organisational cultures as possible. It is not that the companies need to be alike; rather, they need to be aware of their differences, in order to make a strategy of how to deal with these differences.

When wanting to prevent conflicts, as a project manager, or as another person involved in the project, it is essential to get to know as much as possible about the other company; also "beneath the surface". What are their values, how do they internally reward or "punish" subordinates – and thus, what kind of reinforcement does the person risk to get, with whom you are negotiating (or fighting)? This might get you an understanding of what is at stake for the other party.

At the same time, try to get as close a relationship as possible with the persons you are going to cooperate with. Get to know who they are as persons, how do they communicate, what are their personal values, etc.

All in all, this information will be essential for you, when building up a good partnership – but also when having to prevent conflicts from escalating.

To sum up: creating a good partnership should not be left for something you do after working hours, at a less formal business dinner. Neither should it be based on more subjective emotional values: do I like the person from the other company, who I am going to cooperate with (and thus: do I want to spend time with Y?) or don't I like him/her? – and thus: do I prefer to eat in my room alone, instead of having a business meeting?

Creating a good partnership with the external company may be the first and foremost important task of the whole project, and should as such be taken very seriously.

Formalizing the creation of a good partnership is crucial, and a lot of resources should be dedicated to this important preparation of the whole project.

Steps towards a formalized good partnership:

1. The two companies agree on creating a partnership based on a common strategy (common goals for the project) and common strategy for, how to reach these goals.
2. The companies (or a third, external party) make an analysis of the organisational cultures of each company, and an analysis of each person who will be part of the project.
3. Presentation of differences and similarities in the two different organisational cultures, including pointing out potential conflict areas.
4. Leaders of the two companies work out and agree on a common strategy, based on the insight and understanding of their cultural differences and similarities, which will be communicated to all levels of subordinates working on the project.
5. A regular follow-up on results of the project, cross cultural challenges and reminder of how to execute the negotiated strategy.

More information and a computer based program for how to work on the above mentioned steps can be found through L.N. Marguerre (Marguerre 2000).

4.7 Suggestions About How to Work out a Psychological Contract

We suggest two very different tools linked to the area of preventing a conflict from escalating – or to resolve the conflict once it has occurred. A quite formal part of a contract is the so-called "ADR-clause". A less formal way is the internal psychological checklist. Reference is also made to Chap. 6: Litigation and mediation contributing to conflict settlement and Chap. 7: Expertise contributing to conflict resolution

4.7.1 The "ADR-Clause"

An "ADR-Clause"; (Alternative Dispute Resolution-Clause), covers all the alternatives for solving a conflict through the intervention of one or more outside persons. Most of the time, when the term ADR is used, what is meant is, negotiation, mediation or usage of a Referee as facilitator for negotiations. Some think that mediation should no longer be named "alternative dispute resolution", but rather be named "appropriate dispute resolution", since mediation is a tool, which can be used at an earlier stage of conflict escalation than a court trial.

The usage of ADR is based on a deliberate decision of the parties and is as such not something forced upon one of the parties – as a trial would be. ADR should be arranged by the parties involved in the conflict. What is essential here, is that a deal about using ADR should be made beforehand, before a conflict arises – and thus ADR is a preventing tool, as well as it can be a tool for resolution. It is too late to make agreements about ADR once the conflict has occurred – because an ADR-agreement is very unlikely to be concluded, once a conflict is ongoing. An ADR-clause is recommended to be included in the main contract.

We advise that the ADR-clause should be chosen from one of the existing organizations, which offer ADR-services (such as the ICC in Paris or others). Furthermore we suggest that the following aspects should be taken into account in the ADR-clause:

- Each company has the right to call for a mediation.
- Before mediation is called, Top Management of each party should meet and try to solve the conflict.
- If the Top Management does not find an agreement, mediation should start without any delay.
- In the contract it should be noted, whether one or three mediators should be called upon.
- Provision should be made in the contract, how long the mediation may last and after which time-span the parties have the right to go for arbitration or to a state court.
- The contract should ask both parties to keep working on the project despite mediation or legal proceedings might be going on.

4.7.2 Less Formal – the Internal Psychological Check List

As a tool for the Project Manager, we recommend to go through the following Check List before going to call for Mediation or for Legal Proceedings:

- Personality analysis of the Project Manager, and the persons he has to deal with, using the DISC dimensions described in Sect. 4.4.2.
- Analysis of the strengths and weaknesses of the persons involved, why has the conflict occurred? (Eventually name a Monitor of Litigation, see Sect. 6.1.2)
- Analysis of personal values – and differences between the persons.
- Analysis of organizational differences, values, etc., as described in this chapter.
- Analysis of the organizational cultural background; standards and values of the two parties.
- Definition of a strategy to stop the conflict and discussion with the other party.
- What is the problem of the other party, not understood by the own party?
- What is the own party's problem, which is not understood by the other party?

Redefinition of the strategy to resolve the conflict and discussion with the other party. At least three iterative runs should be made through this procedure before calling for mediation or legal proceedings.

4.8 Conclusion to Chapter 4

With the presented model of conflict escalation and the themes in this chapter, three strategies of coping with conflicts from a psychological approach have been presented:

1. Prevention of conflict escalation through better Communication, improved analysis of the people involved and through the understanding of the other party's organisational culture.
2. Professional Handling of the Conflict through improved communication, creation of dialogue and fostering mutual understanding of the parties by elaborating common targets.
3. Resolution of the conflict, if steps 1 and 2 have not solved the conflict a mediation procedure has to be started.

We began this chapter by encouraging project managers to become more effective in their task by becoming competent in dealing with conflicts. As seen in the case study of Sect. 4.1, some of the heavy organizational consequences associated with poorly managed conflicts are wasted time, lowered morale of the personnel, increased workload, lawsuits and costs for all parties. In the following sections three benefits of constructively managed conflicts have been discussed and models of how to avoid or how to successfully handle conflicts have been shown. The main elements of these models were shown as improved communication, open information sharing, higher quality decision making, and improved working relationships.

4.9 Questions on Chapter 4

1. Try to identify your own language when in conflict, or when you feel threatened. Do you have a tendency to communicate in destructive ways? How can you improve?
2. Name elements of a constructive conflict communication.
3. Name elements of a non-constructive conflict communication.
4. Characterize a person who has a high level of dominance.
5. Give attributes of persons with a low level of intensity.
6. Describe a person of high detail orientation.
7. Give elements of how to analyse organisational differences between two companies acting together in a project.
8. What means ADR and which different approaches exist in ADR?

References

Allport G W (1961) Pattern and growth in personality. Holt, Rinehart & Winston, Inc., New York

Allred K (2000) Anger and Retaliation in Conflict: The Roleof Attribution: In Deutsch M, Coleman P (eds) The Handbook of Conflict Resolution. Jossey-Bass, San Fransisco

Borbye E (2001) Hvorfor er du så anderledes? – Jungs typologi i teori og praksis, Dansk Management Forum.

Cattell R B (1965) The scientific analysis of personality, MD: Penguin, Baltimore.

Dana D (2001) Conflict Resolution, McGraw-Hill, Sydney.

Dickson A (2004) Difficult Conversations, Judy Piatkus (Publishers) Limited.

Fitzduff M & Stout C E (2006) The Psychology of resolving global Conflicts: From War to Peace – vol. 1–3 edt., Praeger Security International, Contemptorary Psychology.
Glasl F (1999) Konfliktmanagement. Ein Handbuch für Führungskräfte, Beraterinnen und Berater. 6., erg. Aufl. Bern – Stuttgart.
Hammerich E (2001) Meeting Conflicts Mindfully. Danish Center for Conflict Resolution.
Hogan R (1991) Personality and personality measurement. In: Dunnette M C, Hough L M (Eds) Handbook of Industrial and Organizational Psychology.
Mayer B (2000) The Dynamics of Conflict Resolution. Jossey–Bass, San Francisco.
Marguerre L N (2000) Ph.D.-Thesis on Supply Chain Management, Business School, Copenhagen.
Monberg T (2005) Konfliktens Redskaber. Børsens Forlag.
Kellett P M (2007) Conflict Dialogue – working with layers of meaning for productive Relationships, Sage Publications.
Rosenberg M (1998) Nonviolent Commuication, PuddleDancer Press.
Weiss J, Hughes J (2005) Want Collaboration? Accept – and Actively Manage Conflict Harward Business Review, pp. 93–101.

5 Negotiations Leading to Conflict Resolution

Abstract. This Chapter deals with commercial negotiations as the preferred method to resolve contractual disagreements and other inter-party conflicts in international projects. It analyses the conflict causes and sources, the conflict prevention strategies and countermeasures before it describes the negotiation process with special regards to project disagreements supported by checklists and case studies. It also deals with preparations of settlement negotiation, handling of negotiation break down and negotiations parallel to litigation or arbitration.

Furthermore it deals with the negotiation of delays and possible extension of contract time as a special and difficult case. The negotiation solution versus the litigation solution is illustrated further by 5 more real case studies (as the first one has already been presented in Chap. 4: Preventing Conflicts by Application of Psychology) in order to back up the strong recommendation to negotiate a solution to a disagreement or a conflict.

Key words: Contractual conflicts; Negotiations; "give and take"; Settlement; Conflict source; negative surprises; "trench war"; interpretation; claims evaluation; cost/benefit analysis; negotiation team; "take initiative to first meeting"; attitude; behaviour; settlement agreement drafting; negotiation breakdown; parallel negotiation; settlement execution, delays; time extension; project case studies; lessons learned; world regional differences; style and formalities; personal relations; conflict prevention; conflict stages; resolution phases; interpretation checklist, claims evaluation

5.1 Introduction

International project implementation is often met by a number of adjustments of the scope of work, of the distribution of work, of specification of civil/building works and of equipment/installation. These changes can be caused by:

- Correction of mistakes made in the preliminary design by the Owner and not revealed by the Main Process Supplier/Main Contractor
- Correction of mistakes made in the design by the Main Process Supplier/Main Contractor and not revealed by the Owner (or his consultant) during the approval process
- Lack of sufficient coordination between the civil and building design on one side and the process and equipment design on the other side
- Changes caused by late approval and conditions from Authorities. Reference is made to Sect. 3.3 "Regulatory Obligations"
- Other unforeseen changes or mistakes

Fig. 5.1. Civil Construction of a Cement Plant

The parties in international technical projects are the Owner, his consultant, the operator or end user, the design engineers, the Contractors, the Process Suppliers and the Authorities etc. Obviously they have different interests and different business strategies, policies and cultures. Reference is made to Chap. 2: "Parties, Roles and Interests in International Projects". Therefore the changes mentioned above can easily result in disagreements which can evolve into conflicts in steps or stages as described in Schedule 5.1.

5.2 Conflict Causes and Sources

Conflicts in major supply or turn-key projects are typically caused by one of the following three main factors:

Schedule 5.1. Checklist of Stages in the Development of Project Conflicts

1. Disagreement between the contract parties regarding execution, scope of work, specification, functionality, capacity, time schedule or payments etc.
2. The parties fail to keep the disagreement at an informal level
3. The parties exchange their positions (in writing) regarding the disagreement
4. Possible start of negotiations in order to reach a settlement after exchange of formal positions
5. First chance of negotiated settlement is unsuccessful – none of the parties have sufficient courage to take the initiative to open real "give and take" negotiations
6. Further exchange of supporting arguments for each party's position may be on a higher organizational level
7. The parties are still in contact regarding the issue/conflict but mostly in form of exchange of supporting arguments for each party's own position
8. Second chance of negotiated settlement is unsuccessful
9. The conflict escalates and it affects other areas of cooperation which are basically without any major disagreement
10. The parties are still in contact regarding the issue/conflict but mostly in form of exchange of supporting arguments for each party's position (stronger rhetoric in order to force a solution through)
11. Third chance of negotiated settlement is unsuccessful
12. The conflict escalates further and now it affects all areas of cooperation
13. Now there is a full fledged conflict between the parties

A. The contract specification is insufficient to solve a problem of the scope of work or the distribution of work between the parties.
B. Lack of ability (or will) of one party (or both parties) to handle changes in the scope of work.
C. Unforeseen negative financial impact on one party (or both parties) by the contract execution or by external factors leading to the inability of one party (or both parties) to perform its (their) obligations.

A more comprehensive list of disagreement causes is given in Schedule 5.2.

Schedule 5.2. Causes of Disagreements

1. Misinterpretation of the contract e.g. one party strongly believes to be 100% right in a scope of work disagreement
2. Misjudgment of one's own claims e.g. 3–5 times too high claims amount
3. Lack of one party's ability to adapt to a new project situation or changed circumstances e.g. new regulatory requirements
4. Lack of ability or willingness to cooperate with the other party e.g. writing e-mails instead of negotiating
5. Weak financial situation of one or more of the contract parties e.g. cash flow problems)
6. Weak managerial situation of one or both contract parties
7. Too much sales, engineering or accounting influence on project management

Fig. 5.2. Exterior of Feed Mill Plant in Eastern Europe

Examples of conflict sources are:

1. Change of scope of work and disagreement on whether a certain task is included in the original scope or whether it is a genuine contract variation, justifying a price increase.
2. Unclear distribution of the scope of work between the client and the contractor and neither of them is willing to perform the work in question.

Fig. 5.3. Mechanical Erection of a Cement Plant

3. Disagreement regarding prices necessary to compensate a variation in the scope of work or the change in quantities because of lack of a specified unit price or an unclear unit price.
4. Change of time schedule due to an external influence on one party. Shall that party be compensated and how? If the party affected refuses to act, before an acceptable compensation is agreed upon, how can the "deadlock" be solved?
5. Delay of the supply, of the work or services by one of the parties.
6. Budget overrun and lack of resources due to financial difficulties.

A comprehensive list of disagreement areas can be found in Schedule 5.3 below.

Schedule 5.3. Project Disagreement Checklist

1. Disagreement between the contract parties regarding scope of work, specification, and functionality
   - Variation order disagreement as the Contractor claims, that it is extra work and the Owner considers it as within the original scope of work
   - Stop of work until agreement on the above subject
   - Non-approval of functionality in design or in installation/commissioning
2. Disagreement between the contract parties regarding delays
   - Contractor delay
   - Client delay
   - Delay caused by external condition
3. Disagreement between the contract parties regarding prices, financial guarantees and/or payments
   - Unit price disagreements
   - Measured quantity disagreement
   - Payment delay on the client side
   - Financial difficulties leading to an unreasonable call on a financial guarantee
4. Disagreement between the contract parties regarding basic project conditions
   - Plant location
   - Plant lay-out
   - Raw material specification
   - Process choice
   - Final product quantity and/or quality
   - Environmental aspect
5. Disagreement between the contract parties regarding interpretation/application of local law, regulation and standard
   - Taxes and duties
   - Technical standard
   - Work permit for expatriates
   - Environmental regulation
6. Multiple type conflicts
   - Financial problem for one party limiting his ability/willingness to find compromises on technical/scope of work disagreement
   - Technical/scope of work disagreement combined with delay and disagreement on contract interpretation

Schedule 5.4. Project Conflict Prevention Checklist

1. Describe the disagreement between the contract parties in afew words.
2. Describe the other party's official position in a few words.
3. Describe the other party's informal position in a few words.
4. Describe our party's official position in a few words.
5. Describe our party's informal position in a few words.
6. Why has the disagreement not been settled already?
7. When have the parties last discussed the disagreement and what was the outcome or action agreed?
8. What is the gap between the parties' official positions?
9. What is the gap between the parties' informal positions?
10. What can the other party do to reduce the gap between the informal positions?
11. What can our party do to reduce the gap between the informal positions?
12. The other party's chief representative and decision maker – do we know who she/he currently is?
13. Has our party appointed a chief representative?
14. Do the two chief representatives know each other from previous business transactions?
15. When has there last been a serious contact between the parties on the disagreement in question?
16. Why has our party not made any approach to the other party in order to establish negotiations?

In settling these disagreements it has to be remembered that there is an important difference between a "normal business deal" and a contractual settlement. The business deal is characterized by the choice to close the deal or to walk away from it. The contractual dispute settlement has to be made out of shear necessity. Here, there is no option to walk away. Senior Management and the Project Managers should be fully aware of this fact.

The checklist shown in Schedule 5.4 "Project Conflict Prevention Checklist" is made and presented here in order to help the reader to prevent conflicts by solving disagreements before they become a conflict.

5.3 Why Commercial Negotiation is the Preferred Method of Conflict Prevention

Even if the project and its distribution of work is better defined and the parties are better prepared and even if the contract is better drafted there still will be changes that give rise to disagreements between the parties. In any project there will be unforeseeable changes that require to be handled by the parties in a commercial manner. Furthermore the parties have opposite financial and managerial interests. Therefore disagreements will occur and the key issue of this book is how to handle and settle these disagreements in an effective way so the project gets completed on time and with a satisfactory result for the parties.

Basically there are the following 3 methods available for reaching a negotiated settlement:

A. "The wait and see approach" where the parties choose to postpone the settlement until the project is finished and the final accounts between the parties are to be made and agreed.
B. "The tactical approach" where one party (the one that feels stronger) forces the other weaker party to agree to the first party's position (more or less). The weaker party does not at all like the solution, but can't resist the pressure!
C. "The strategic and pragmatic approach" where both parties accept, that their full cooperation is needed to complete the project work and accept that adjustments to the original plan, the contract, schedule and the budget are needed due to unforeseeable changes in circumstances, mostly beyond the control of the parties. In order to make such adjustments the full cooperation requires that disagreements are resolved as the project makes progress.

We are definitely in favour of the strategic and pragmatic approach as a principle. This method should not be misunderstood as a weak and soft attitude. It does not at all rule out tactical manoeuvres to reach a commercial negotiated settlement. In this context there might be situations where one party has to "make a point" and stand firm on certain facts and his interpretation of the disagreement. But the solution is a genuine commercial settlement acceptable to both parties.

Method A, in our opinion, does not work as a general principle because:

- Balance of power between the parties shift drastically when the work has been completed.
- Notification and the factual description of the case have anyway to be done right after the occurrence and can not wait till the project is completed.

Method B – the disadvantages and risks of this method are the following:

- It might work in one case but not in all the cases during project implementation for obvious reasons (the weaker party learns by experience).
- If used, it can backfire in the next situation and result in a conflict that is difficult to handle (the weaker party might take revenge on his loss).

Method C is the preferred method because it is the most cost effective, the fastest and the best for the business. But this conclusion can only be made under the assumption that the outcome by litigation is not significantly better than the commercially negotiated settlement outcome.

Table 5.1 shows a typical example of a cost/benefit analysis comparing a commercially negotiated settlement method with a litigation method in a disagreement between a Main Contractor (Us) and a Subcontractor (Them). As can be seen in the case the commercially negotiated settlement method is the fastest and most cost effective as is very often the case in the real project world. This of course has to be proven in the actual case.

Table 5.1. Cost-benefit analysis of claims situation (from Main Contractor's position)

Category	Status / Position	Their amount €1.000	Worst case evaluation €1.000	Medium case evaluation €1.000	Best case evaluation €1.000	Our amount €1.000
1. Contract value adjusted						
1.1 Original contract value	Signed & effective (S&E)	2500	2500	2500	2500	2500
1.2 Extra order A	(S&E)	34	34	34	34	34
1.3 Extra order B	(S&E)	52	52	52	52	52
1.4 Extra order C	(S&E)	45	45	30	16	16
Total revised contract amount		2631	2631	2616	2602	2602
2. Sub-contractor's claims (Them)						
2.1 Claim 1	Notified Date (ND)	159	159	95	40	40
2.2 Claim 2	(ND)	185	185	100	0	0
2.3 Claim 3	(ND)	26	26	20	15	15
2.4 Claim 4	(ND)	97	97	40	18	18
Total claims against us		467	467	255	73	73
Total revised contract amount minus their claims		3098	3098	2871	2675	2675
3. Main contractor's claims (Us)						
3.1 Claim 5	(ND)	20	20	75	80	80
3.2 Claim 6	(ND)	92	92	105	105	105
3.3 Claim 7	(ND)	67	67	75	80	80
3.4 Claim 8	(ND)	98	98	109	119	119
Our total claims against them		277	277	364	384	384
Net claims – Main Con's viewpoint		(190)	(190)	109	311	311
Total revised contract amount minus net claims before costs		2.821	2.821	2.507	2.291	2.291
4. Our cost of claims settlement by negotiations						
4.0 Expected duration in months	PM guestimate (PMg)		6	4	3	
4.1 Preparation costs (internal)	(PMg)		10	8	6	
4.2 Travelling and meeting costs	(PMg)		10	8	6	
4.3 Cost of meeting time	(PMg)		10	9	8	
4.4 Settlement assistance	(PMg)		12	10	8	
4.5 Total settlement costs	(PMg)		42	35	28	
4.6 Expected outcome excl. interests	(PMg)		(190)	109	311	

Table 5.1. Continued

4.7 Expected net outcome (outcome – costs)	(PMg)	(232)	74	283
5. Our cost of claims settlement by litigation				
5.0 Expected duration in month	Legal Dept. gues-timate (LDg)	22	18	14
5.1 Preparation costs (internal)	(LDg)	30	24	14
5.2 Travelling and meeting costs	(LDg)	40	30	12
5.3 Cost of own time assist-ing/managing lawyers	(LDg)	30	25	6
5.4 Legal assistance	(LDg)	120	100	30
5.5 Court fees and costs	(LDg)	50	40	20
5.6 Total litigation costs	(LDg)	270	219	82
5.7 Expected outcome incl. in-terests	(LDg)	(190)	109	311
5.8 Expected net outcome	(LDg)	(460)	(110)	229
6.1 Net position after settlement	(LDg)	(232)	74	283
6.2 Net position after litigation	(LDg)	(460)	(110)	229
6.3 Difference in net position	(LDg)	228	184	54

Note: Fictional case based on real experiences

Often this basic difference is not realized by the parties of a contractual conflict and they believe that "walking away without a deal" is an option. In a project such deals have to be made and the work has to continue!

The key issue is, why Top Management does not react in a much more pragmatic way to avoid conflicts by settling the differences before they become a conflict! The origin of business conflicts involves human aspects that are dealt with in Chapter 4. Here we can only state the fact that humans sometimes get into conflicts because of different interests, systems, attitudes, understanding, culture and communication.

Negative surprises which are difficult to handle are definitely also a contributing factor to conflicts in projects. It is a general experience, that problems between two parties are often caused by surprises such as financial deviations, which in a project tend to spread to other parties via conflicts.

5.4 How Can the Rate of Success in Commercial Negotiations Be Improved?

We have all witnessed commercial negotiations where one or both parties really were not interested in reaching any negotiated agreement. The party/parties involved

Fig. 5.4. Completed Cement Plant in Operation

are often badly prepared or not at all prepared, such that their attitude is non-constructive, bureaucratic and un-committed and therefore the commercial negotiation fails to produce any acceptable agreement. After some time and several attempts the commercial negotiations are given up by both parties (here they can agree!) and the case is referred to litigation or arbitration (to the relief of both parties, who do not dare to propose a compromise, which makes the other party interested in coming back to the negotiation table).

In our opinion mainly 3 factors are decisive for the success of commercial negotiations:

1. The project culture of the company, its project management style and behavior, reference is made to Chap. 8 "Project Management".
2. The competences, training and personal (i.e. his personal qualifications) qualities of the key staff in charge of the project (notably the Project Manager and the Construction Site Manager) and the Senior Manager in charge of the negotiations – reference is made to Chap. 4. "Preventing Conflicts by Application of Psychology" and Chap. 6. "Litigation, Arbitration and Mediation Contributing to Conflict Settlement" (the Monitor of Litigation role that can very well be applied to commercial negotiations as well as litigation).
3. Use of conflict prevention routines in daily project work and especially on the construction site:
   - Weekly site meetings with minutes of meeting and a standard agenda that includes disagreements, as for instance work orders requested but not approved
   - Agreed format and procedure for daily and monthly progress report that includes disagreements
   - Agreed format and procedure for the exchange and follow up on claims and other disagreements

- Formal and informal contacts between the parties on a daily basis
- "Hotline system" when a disagreement threatens to evolve into a conflict

5.5 The Contract Parties and Their Situation

Firstly an early detection of a potential project conflict is very important thereby avoiding the "frozen positions" and the disagreement developing into a "trench war", where the "grenades" are arguing mails, faxes and letters. Here reference is made to Schedule 5.1: "Checklist of Stages in the Development of Project Conflicts".

Secondly it is recommended that each party, carefully and as objectively as possible, analyses at first the situation of the other party and then its own.

The following factors should be taken into account in such an analysis:

A. The parties' contractual positions (what is right and what is wrong)
B. Claims estimation and documentation of purpose, quantity, prices and costs
C. The financial situation of the other party
D. The management situation of the other party
E. The project time schedule, organisation, resources and competences

Third parties that might influence the parties in a strong way, such as:

- Financial Institution involved
- Other third party with influence on the operational situation (e.g. end user, inspection authority, consulting engineer)

5.6 Preparation of Negotiations

5.6.1 Basics in Preparation of Negotiations

It is often said that the success of a settlement negotiation "mostly depends on the attitude of the other party". We do not agree to this principle – to the contrary! In our opinion the prime success factor is one's own preparation before the negotiations start. An important part of one's own preparation is the evaluation of the position of the other party. This leads to an analysis of possible settlement options instead of requesting immediately arbitration or court procedure.

Schedule 5.5: "Contract interpretation checklist part 1 & 2" and Schedule 5.6: "Checklist for claims evaluation" are examples of how such an analysis can be formatted.

For each claim forming part of the case it is a good idea to state the facts of events, notifications, contractual justifications and calculation of claimed amounts in a systematic way with attached documentation. During negotiations a presentation of such an analysis will allow the other party to accept, modify or decline each element, so his position and his strength can easily be understood and evaluated by the first party. Table 5.2: "Analysis of contractual claims situation" is an example of a schedule for a systematic way of analysing the "pros and cons" of a settlement versus litigation.

Schedule 5.5. Contract Interpretation Checklist, Part 1

1. Make sure you have a genuine copy of the full original contract incl. all annexes and other documents/standards referred to or applicable.
2. Provide and read copies (signed documents) of all relevant contract documents.
3. Identify and list all relevant contract provisions applicable, specifically and generally.
4. List the order of priorities and decide solution of conflicting provisions.
5. Can the supplier provide the contractual proof that the actual equipment spec. or scope of work is, what is necessary to fulfill the contract?
6. Write down the conclusion regarding, who carries the burden of proof.
7. Write down supplementary questions and conditions.
8. Check the conclusion and conditions with a colleague not involved previously.
9. Consult with Legal Department and/or External Lawyer in critical and major cases.
10. Write down a case study in a very short form as an "Aide Memoire" that might be handed over to the other party.
11. Analyze the other party's contract interpretation position documented as far as possible.

Schedule 5.5. Contract Interpretation Checklist, Part 2

Special considerations regarding scope of work, limit of supply and spec disagreement

A. Disagreement regarding the scope of work:
   1. Contract type and general provisions regarding scope of supply or works
      - Turn-key or complete supply contract with overall responsibility by supplier or contractor
      - Fit for the purpose provision
      - Single machine order or similar restricted responsibility
   2. Method for determination of scope of supply or works
      - Scope of work description and Equipment Lists
      - List of Client's supply and obligations
      - Division of Responsibility and Border Line Activity List
      - Battery Limit lay-out/arrangement drawing
      - Verbal description of limit of supply supplemented by drawing (e.g. "at outlet flange of Pump no. 765")
B. Disagreement regarding specifications:
   1. Contract type and general provisions regarding scope of supply or works (as above)
   2. Method for determination of contractual specification
      - Dimensions, materials, capacity and other characteristics
      - Process Description and Performance Data
      - Norms & Standards
      - Type and Manufacturer
      - Watch out for conflicting specification – in some contracts the solution is to supply in such a way that all specs are fulfilled!

Why do we recommend to evaluate the case as seen from the point of view of the other party? The short answer is that such an evaluation will make our own assessment much more realistic with respect to what can be obtained by a negotiated

Schedule 5.6. Claims evaluation checklist

1. What happened, that caused the claim and who was present (factual description of incident(s))
2. Opponent's full claims amount (preliminary or final amount)
3. Opponent's claims notification* (letter/e-mail and other documentation attached)
4. Does the notification contain a reasonable description of possible consequences?*
5. Date of incident(s), date of notification and date of claims*
6. Documentation of consequences causing the claim*
7. Contractual justifications – can we prove that the party supposed to be "guilty" has a contractual responsibility?* (according to the evidence of the incidence and according to the contract)
8. Event causing consequences incl. other events/consequences (is there a direct cause-effect link between the cause and the alleged effects and were there other causes for the effects?)*
9. Estimation of physical consequences*
10. Estimation of cost of consequences (extra costs only) – are they fair and reasonable?*
11. Evaluation of main claims factors such as claims notification, justification, consequences and estimation of extra costs
12. Opponents expected final amount claimed versus our objective evaluation

Note: Elements marked with * are very critical for the claims evaluation

settlement or by litigation. Furthermore it might remove most of the emotional factors influencing the evaluation.

What can be done to make such an evaluation realistic? The short answer is to involve internal or external advisers with no history or stakes in the project adds to the objective analysis. Reference is made to Chap. 6 about mediation and Chap. 7 about third party expertise.

It is our experience that parties in many cases overestimate their own net outcome position by a factor from 2 to 5, which of course is grossly misleading and very impractical from a general management point of view. The use of systematic evaluation and neutral advisers will lead to a more realistic estimation of the outcome and increase the chances for a reasonable settlement.

5.6.2 Negotiating Team

It seems obvious that the negotiating team also should consist of staff not previously been involved in the dispute in order to disregard emotional aspects. This principle is unfortunately often compromised and normally the teams mostly consist of the project manager, the site manager and other project staff already involved. Sometimes the delegation is headed by the project director. He might not have been so much involved earlier, but he has a major stake in the short term outcome.

Instead, staff with experience in settlement and ability to obtain the confidence of the other party should also participate. If already appointed, the "Monitor of Litigation" (see Chap. 6) could lead such negotiations.

Table 5.2. Analysis of the Parties' Positions (from Subcontractor's position)

CATEGORY	Their official claims amount € 1.000	Their last settlement proposal € 1.000	Our estimate of a possible settlement amount € 1.000	Our last settlement proposal € 1.000	Our official claims amount € 1.000
1. Contract value revised					
1.1 Original contract value	2500	2500	2500	2500	2500
1.2 Extra order A	34	34	34	34	34
1.3 Extra order B	52	52	52	52	52
1.4 Extra order C	45	45	30	16	16
Total revised contract amount	2631	2631	2616	2602	2602
2. Subcontractor's claims (Them)					
2.1 Claim 1	159	159	95	40	40
2.2 Claim 2	185	185	100	0	0
2.3 Claim 3	126	126	120	115	115
2.4 Claim 4	197	197	140	118	118
Total claims against us	667	667	455	273	273
Total revised contract amount minus claims	3298	3298	3071	2875	2875
3. Main Contractor's claim (Us)					
3.1 Claim 5	20	20	70	74	80
3.2 Claim 6	92	92	100	100	105
3.3 Claim 7	67	67	70	75	80
3.4 Claim 8	98	98	105	109	119
Total our claims	277	277	345	358	384
Net claims – SubCon's viewpoint	390	390	110	– 85	– 111
Total revised contract amount minus net claims before costs	3021	3021	2726	2517	2491

Another possibility is to establish a new self contained and powerful team fully able to negotiate and settle according to an agreed mandate from Top Management assisted by the Project Manager etc.

In our experience, nothing is gained by a protracted or interrupted settlement negotiation, because one party can not manage the process and its elements professionally. It is definitely an advantage that the team is strong and self contained with respect to documentation, calculation and evaluation of legal aspects and has the power to agree and draw up the settlement agreement. But even so, it is normal that a telephone call to Top Management is necessary to confirm the final agreement and its signing. But the power to negotiate should be with the negotiating team and notably its leader and not with someone at the head office.

Each party of course decides its own team but the initiator of settlement discussions should propose the organisational level of the team leaders and the maximum number of participants from each side. But more important than levels are commitment and availability.

5.7 Suggestions on How to Start Settlement Negotiations

The first and often asked question is, who should take the first initiative and propose settlement negotiations? If there are regular contacts between the parties at all levels, this issue is "none existent" as the idea of negotiating a settlement would come naturally without anyone noticing who initiated it. But as daily contacts are often scarce, it is in practise a crucial issue and it could also be a matter of pride.

The parties might consider it to be a sign of weakness to take the initiative and propose negotiations. In our opinion the contrary is true. An invitation to negotiate can well be considered as a sign of strength since it demonstrates professionalism.

Another positive aspect of being the party taking the initiative to start negotiations is the influence on the first meeting place, agenda and participants. This should not be used to manipulate but to make a good and productive start to the negotiations.

It is also a very good idea to agree on the agenda for the meeting which could be used as a basis for short factual and action orientated minutes of meeting. Our experience shows that such neutral and factual Minutes of the Meeting signed by the parties facilitate the negotiations and the preparation of a written settlement, because they gradually build up confidence between the parties during negotiations. Secondly Minutes of the Meeting as an integrated part of the meeting facilitates systematic negotiations, and thereby makes them more efficient.

Generally it is recommended to be pragmatic, flexible and compromising regarding these practicalities creating a good negotiating atmosphere and demonstrating the will to create a settlement. Reference is made to Schedule 5.7 "Conflict Resolution Phases Checklist".

Schedule 5.7. Conflict Resolution Phases Checklist

1. Identification of a serious disagreement or an upcoming conflict between contract partners in a project
2. Classification of the serious disagreement by type, scope and possible consequences
3. Decision regarding the management priority and the degree of urgency in solving the conflict
4. Description of other party's claim and of his position and analysis of possible solutions and risks
5. Arrange clarification meeting with the other party to discuss the conflict and its possible solution
6. Develop a certain trust between the parties e.g. by discussing the technical aspects
7. Cost/benefit analysis of possible solutions
8. Choose the priorities of possible solutions
9. Planning of the implementation of the best solution
10. Decision on which solution to implement and establishing a negotiating mandate
11. Appoint a chief negotiator and a team
12. Negotiations until a settlement is reached incl. adjustment of the mandate
13. Settlement implementation and registration of consequences
14. Experience collection and feed back

One example is the meeting place which might be very important for some companies. Therefore we suggest letting the non-initiating party choose the meeting place for the first meeting and to agree then to rotate the meeting place.

5.8 Negotiations Leading to Settlement

5.8.1 Basics of Negotiations

For negotiations to be successful the communication between the parties must work satisfactorily. The parties must be able to get their points and facts across to the other party in a friendly and understandable way. There are many examples of miscommunication. Once at the end of a negotiation session a European main contractor was proposed a settlement solution by a subcontractor in a very short, fast and quite arrogant way. As a result the Main Contractor failed to understand that it was a serious settlement proposal to be accepted or refused. The communication failure was only revealed later and therefore the settlement attempt failed completely.

Trust is another important factor. A conflict often originates from mistrust. Negotiations need a certain trust level between the parties at least in respect of honest proposals, procedure, agenda, participants, time and place. Unwarranted changes in proposals, planning, location and participants of a meeting can have very negative effects on the chances of success. Trust in technical and other factual information can be very beneficial for the negotiations and will pave the way to a successful settlement. Furthermore trust can be built up during negotiations. One way of building up trust is to produce reliable and fair minutes of the meeting and get them signed by both parties at the end of each meeting. The preparation and signing of Minutes of the Meeting is a good training in the drafting and eventually signing of the settlement agreement.

The first meeting is especially important. Reference is made to Schedule 5.8 "Agenda Checklist for First Negotiation Meeting".

Informal contacts in between the meetings can enhance communication and trust building. The informal contact is recommended to pass through a coordinator on each side. Such coordinators should act as secretaries to the negotiating teams and are responsible for progress, practicalities and preparation. The team leader or chief negotiator is the representative of the party and the team leader presents the position of the party and makes decisions on what to agree on or to recommend for approval by Senior Management. There should always be a higher management level back home at head office to refer to. If this is not the case, there is a risk of a "lose–lose" situation (the opponent takes advantage of the active participation of a Senior Manager, who can not return without a result).

Internal preparation and follow up on negotiations are very important. Table 5.2: "Analysis of parties' positions" can be used to follow up on the development of the positions of the parties. In a major contract the final account can be so complicated that systematic calculation in a spread sheet is necessary. The systematic and terminology can very well be shared by both parties.

Schedule 5.8. Agenda Checklist for First Negotiation Meeting

1. Purpose of meeting:
   - Clarification meeting
   - Settlement investigation meeting
   - Negotiation meeting
2. Meeting formalities proposed and agreed:
   - Agenda & Minutes of Meeting
   - Participants and team leaders
   - Meeting frequency, schedule and places
3. Establishing the factual background for the claim and disagreement
   - Background
   - Agreed facts about what happened
   - Not agreed circumstances of what happened
4. Claimant's position
   - Cause – effect – claim by claim
   - Contractual justification – claim by claim
   - Consequences on the work and time schedule
   - Estimation of extra costs
   - Claimant's proposal and position
5. Defendant's position
   - Cause – effect – claim by claim
   - Contractual justification – claim by claim
   - Consequences on the work and time schedule
   - Estimation of extra costs
   - Defendant's proposal and position
6. Points that need to be investigated further
   - Facts about what happened
   - Cause – effect
   - Contractual justification and interpretation
   - Estimation and documentation of extra costs
7. Other proposals
   - Procedural issues and proposals
   - Material issues and proposals
8. Next meeting
   - Date, Place and time
   - Participants (each party decides their own participants)

5.8.2 Attitudes and Behaviour of Negotiators

In order to obtain good communication and thereby enhance a settlement result, there is a need for the right attitude and behaviour of the negotiators. This requires in our opinion the following seven qualities:

A. Patience in presenting, explaining and arguing the case
B. Ability to "get across the message"
C. Flexibility in finding solutions and seeking compromises

D. Keen to make results by consensus and demonstrate the will to make compromises in order to close an agreement
E. Stamina to continue the work for days until a result is achieved
F. Sensitiveness in understanding the other party's background, arguments, reasoning and position
G. Sufficient command of the negotiating language and diplomatic capabilities

5.9 Making the Agreement for a Settlement

5.9.1 Basics of a Commercial Settlement

A negotiated settlement is by experience much better than a court "confrontation" because it is faster, more likely to be successful and less costly. Finally it is much better for the future business not only between the parties involved but generally. We will come back to the cost aspect. Here we will only mention that normally a negotiated settlement of major contractual issues takes weeks or months to achieve, whereas litigation normally takes years.

Many factors are involved in making a negotiated commercial settlement as described in this chapter, but in our opinion two factors are the most important ones:

A. A careful understanding and analysis of the position and the reasoning of the other party
B. A neutral and objective estimation of the predicted outcome

Very often negotiators are so excited by their own claims position, which they tend to overestimate due to emotional factors. The advantage of negotiations is that they force us to understand and analyse the facts and reasoning put forward by the other party regarding our claims. Thereby they form a more realistic basis for our own evaluation of a likely outcome in a possible litigation.

The negotiation also enables us to influence the other party, where our arguments are strong and vice versa. This will bring the two positions closer to reality and to each other.

5.9.2 Final Settlement Discussions

When all claims have been carefully presented, documented, justified and discussed it is time to map them together in a factual and neutral way and put money on them so that the total amounts involved can be summarized as shown in Table 5.2.

Finally the net amount to be paid by one party to the other has to be agreed and the conditions under which the payment will be made have to be attached. It is recommended to start with agreeing on the conditions assuming a net amount can be agreed upon. Especially the payment conditions need to be perfectly clear and safe.

5.9.3 Drafting the Settlement Agreement

As soon as the parties can envisage that a settlement might be within reach they should start drafting the settlement agreement, so when the final net amount has been agreed together with conditions and actions, it can, without too many problems, be inserted into the agreement text for final scrutiny and signing.

Schedule 5.9 "Checklist for drafting of settlement agreement" gives the main points to include in the draft. It is very important to consider the following questions before the drafting starts. The first question is, whether the settlement agreement is in line with the contract and the conditions and obligations of the agreement fully respect the contract. The second question is whether the settlement agreement can be enforced and executed without further agreements. If these two questions can be confirmed, then the parties can go ahead with the final drafting and signing of the settlement agreement as outlined.

If not, the parties definitely need to consult with their lawyers. That is also to be recommended if any of the settlement obligations are outside the contract scope

Schedule 5.9. Drafting of Settlement Agreement Checklist

1. Reference to the Contract and Parties clearly identified (identical with contract parties)
2. Place and date of settlement agreement and authorized representatives of parties signing incl. full name and title
3. Claims included in a full and final settlement each clearly identified with reference to letters (especially the notification letter), minutes of meetings and to the respective contract sections
4. Claims and issues not included in this settlement agreement
5. Net settlement currency amount in figures and words
6. Payment conditions clearly specified (when, where and how)
7. Consequences of possible late payment of settlement compensation – interest rate specified
8. Conditions for the settlement agreement
   - 8.1 Work to be completed or rectified
   - 8.2 Supplies to be delivered
   - 8.3 Acceptance of supplies, works, services and documentation
   - 8.4 Obligation to supply spare parts and render services incl. conditions
   - 8.5 Delivery schedule
   - 8.6 Warranty obligations
   - 8.7 Release of financial guarantees
   - 8.8 Other conditions
9. Consequences of the settlement agreement
   - 9.1 Global commercial settlement without prejudice to other claims
   - 9.2 Confidentiality
   - 9.3 Payment amount, conditions and execution details
   - 9.4 Acceptance of supplies, works, services and documentation
   - 9.5 Provisions for handling of potential disputes arising from the execution of the settlement agreement incl. governing law and dispute resolution provisions
   - 9.6 Other consequences of the settlement e.g. remaining obligations under the contract

and/or if ongoing or complicated delay situations are involved. But it is still recommended that the negotiations to be lead by the business managers of the parties and not by the lawyers.

5.9.4 Concluding and Signing the Settlement Agreement

The signing of the agreement should be made immediately upon having agreed on the net amount and the conditions of the settlement. In surprisingly many cases the settlement agreement is signed at a later date and this is often causing "some clever guys" to try to introduce new aspects and thereby causing serious and unnecessary problems. In order to achieve a "once through settlement agreement", it is recommended to exchange drafts of the final agreement before the parties meet for the final negotiation and the signing.

The team leaders should prepare themselves for signing of the agreement by making telephone calls to their respective head offices in order to obtain approval or clearance of the signing of the settlement agreement. It weakens the position of the team leader, if he has to come back with changes as instructed by his Top Management, especially if they are of non-principal importance.

All pages should be initialled, including exhibits and attachments which form an integral part of the settlement agreement and should always be signed prior to the time of signing the agreement text itself.

As the purpose of the settlement is to re-establish or improve cooperation between the parties the settlement agreement should be acceptable to both parties and enhance the spirit of cooperation. Therefore payment conditions and other actions should be clear and simple to avoid new disputes.

5.10 Handling Break-down of Negotiations

Settlement negotiations might be stopped or might break down for mainly three reasons:

A. The distance between the positions of the parties is too big to overcome at this point in time
B. One of the parties is not able to formulate its position and/or decide, how to react to the position of the other party (for instance due to managerial and/or financial problems)
C. It is not possible to conduct real negotiations between the parties, because of a lack of confidence, trust or similar problems

Reason A: In this case it is recommended to put the negotiations on hold in a waiting position and resume them when both parties agree that the gap might be closed by more flexibility.

Reason B: This is a difficult situation to solve. It will most likely result in litigation. A special effort has to be made to overcome such a situation.

Reason C: In this case the parties seem to have a need to rebuild trust and confidence from scratch. It can be recommended to split up the issues of the case and to start with minor and simple outstanding issues as for instance faulty delivery of parts or a problem related to the documentation supporting an invoice. When this problem has been solved, the parties can continue by solving other more complicated problems. An external mediator might also be of assistance if the parties can agree on one person they both have confidence in. Or they might agree, each to appoint external representatives with no history in the project and no bad feelings about what happened!

5.11 Negotiations Parallel with Litigation

Parallel negotiations during litigation are not only possible but also strongly recommended. All litigation institutions (e.g. state courts or arbitration organisations like ICC Court of Arbitration) accept and even encourage parallel negotiations between the parties aiming at finding partial or global settlements.

The parties are not obliged to inform the court in charge of the litigation process that the parties have started settlement discussions or negotiations. When a settlement has been concluded, then it is necessary to inform the court (state or arbitration court) in charge to stop the litigation process.

The Judges or the Arbitrators will most likely be pleased that the parties try to settle out of court, the practical, technical or accounting issues, which are outside the expertise of the Judges or Arbitrators. They definitely prefer to deal with the principal legal matters and advise the parties accordingly, enabling them to negotiate details, acceptance of the project, and its final accounts.

Even when the parties agree on a final settlement including some legal issues, the Judges or the Arbitrators will in most cases be happy. As outsiders they are likely to wonder, why the parties have not been able to settle the issues or conflicts before.

The lawyers will recommend the parties to mark all written material handed over to the other party: "Without prejudice to the legal and material position of the parties in the litigation – both parties reserve all their rights".

5.12 Executing the Settlement Agreement

No settlement can be considered as concluded until the agreement has been fully executed i.e. all remaining supply and works have been satisfactorily performed and signed off and all payments and/or release of guarantees have been made.

The effort invested in reaching the settlement should be continued during its execution. Otherwise all the efforts are wasted. Often the follow up on the execution of the settlement agreement after the signature is too weak, resulting in delays and in the worst case in a continued conflict.

Sometimes the settlement is only formalized in a short preliminary agreement signed to be followed and replaced by a "legal correct" version. This dual stage agreement method is not recommended as it often leads to an unnecessary complicated negotiation. Either the lawyers participate in the drafting of the first settlement agreement or they do not participate at all. For most settlement agreements their main role should be to draft or revise the settlement agreement text from a contractual point of view.

Approval and ratification of the agreement by Senior Management after the conclusion of negotiations and immediate signing of the agreement can not be recommended. Our advice is, that the team leaders on both sides have full power to negotiate, draft and sign the agreement and execute it afterwards. It is their obligation and interest to get approval or clearance from Senior Management prior to negotiations, and at the latest before the final round of negotiations including the signature of the settlement agreement.

5.13 Negotiation of Delays and Extension of Time

Previously most contractual conflicts were related to extra supplies or extra work, to specification issues or to limits of supply or work. Nowadays it seems that the emphasis has shifted to delays!

The focus on delays is caused by shortening of delivery time or completion time and the fact that clients are becoming more and more strict on keeping the contractual time schedule, which means that penalties and delay damages are claimed and collected in case of any delay.

In most Engineering, Supply and Construction Contracts the planned and agreed time of delivery (or of completion) is determined by a fixed date. In order to establish at which date, the Supplier or Contractor actually delivered or completed their works, written evidence has to be established, preferably with the signature of both parties (e.g. certificate of substantial completion). If that date is at or before the planned and agreed delivery date, the Contractor has delivered on time; if the date is later, there is a delay, which is considered the Contractor's delay, if he can not prove otherwise.

Contractually the Contractor will in principle bear the responsibility for all delays including delays caused by factors beyond his controls i.e. Force Majeure, delays caused by the Clients, by other Contractors/Suppliers or by local authorities unless the Contractor can prove otherwise!

In order not to become responsible for other parties' delays, the Contractor needs to protect his "contractual delay interests" by acting when the delay occurs.

This action to protect his "contractual delay interests" should in our opinion consist of the following activities:

- Collection of factual evidence that a delay causing factor occurred and for which reason
- Registration of witnessed evidence
- Delay notification to the other contract party including description of expected consequences
- Registration of the delay factor and its consequences – computation of cost simultaneously with the consequences

- Relate delay evidence to contract clauses as justification
- Preparation of a delay file for the other party
- Negotiation of delay consequences in terms of time extension and extra cost
- Settlement of delays simultaneously with project progress instead of waiting until the project is completed

The delay file is recommended to consist of the following sections:

- Description of the delay causing events and the consequences – very factual incl. documentation of evidence
- The Contract provisions that describe the work or supplies influenced and the distribution of obligations and responsibility to the respective parties
- Description of the factual deviations between what was foreseen in the contract and what actually happened
- Analysis of cause – effect relation: Only effects directly caused by delay causing events can be considered and if other factors are involved (e.g. own shortcoming) it will reduce or eliminate the accepted delay effects!
- Estimation of the time extension which should be requested on contractual grounds only (delay directly caused by the event and provided that diligent action has been taken by responsible party)
- Estimation of the additional costs claimed e.g. cost of staffing in the extra period of delay

The volume of such a file can be 20–150 pages including documentation. As the claim preparation work has to be done at the same time as the delay occurs, and simultaneously with project progress, it requires competent staff and resources. If these resources are not available, then the delay file will not be prepared and can not be presented or will be rejected as insufficient and extension of time will not be granted. We recommend to use an outside expert for file preparation if the in-house capacities are not sufficient.

It is crucial to prove the following two relations:

1. That the event causing the delay was not at all within the responsibility or control of the claiming party
2. That there is a direct relation between the event causing the delay and its effect on the work, and that it is really causing extra costs

When more than one delay occurs or has occurred it might be difficult to keep track of the various delays and time extensions claimed. Figure 5.5 based on Microsoft Project 2007 Gantt diagram is a systematic, professional and precise way of presenting the starting and end dates, the durations of the contract period, the delays and the extension claimed. Schedule 5.10 is a short description on how to operate this method.

Even when the delay file is made according to the claim requirements as stated above, it takes a lot of courage to present it to the Client in a serious way, right after the delay has been notified. In most cases the Client would try to postpone the discussion and the negotiation until after completion. But the Contractor's Project Manager has

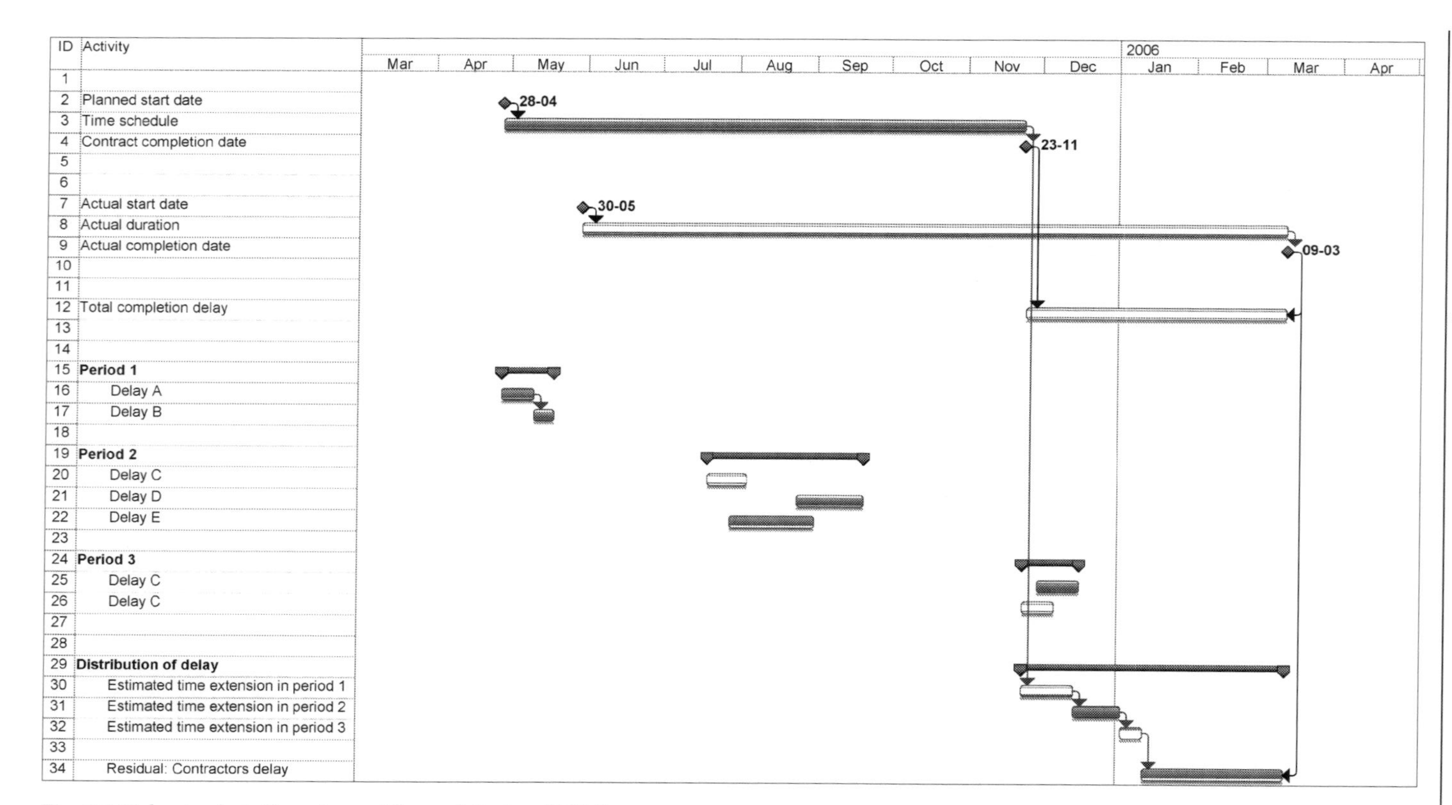

Fig. 5.5. Delay Analysis Overview – Microsoft Project 2007 Gantt

Schedule 5.10. Guidelines for mapping delays in Microsoft Project 2007

1. Contractual time schedule – insert the following data from the contract and the signed amendments:
   1.1 Agreed start date as per effective contract (specific date)
   1.2 Agreed total duration as per effective contract plus signed amendments (in calendar days)
   1.3 Agreed end date as per effective contract plus signed amendments (specific date)
2. Actual schedule
   2.1 Actual start date (specific date)
   2.2 Actual total duration (in calendar days)
   2.3 Actual completion date (effective date of acceptance certificate) (specific date)
3. Mapping of overall delay: Actual completion date (2.3) minus Agreed end date (1.3)
4. Claimed delays
   4.1 Claims for extension of time should be organized in groups each containing claims time wise next to or overlapping each other, so the system can calculate the overall time for each group. This information assists the evaluation.
   4.2 One activity for each claim: claimed start date, duration and end date
5. Extension of time
   5.1 Evaluated likely decided extension of time (settlement between the parties or award by court or arbitration tribunal)
   5.2 Factors to be taken into consideration in the evaluation
      5.2.1 Notification in due time
      5.2.2 Contractual justification
      5.2.3 Proof of start date, duration and end date
      5.2.4 Degree of progress at disturbance or hindrance
6. Contractors delay: Overall delay minus total evaluated likely extension of time

to convince the Client to discuss, negotiate and settle the delay file when the delay occurs or just after, since then his position is much stronger than after completion. The Client will of course be more easily convinced, if he can trust, that the extension of time granted, will not necessarily change the agreed completion date (on a best effort basis). Furthermore it helps, if the Client is convinced that the Contractor's Project Manager is a valuable guarantee for a successful project implementation in all aspects including trying to catch up delays.

From the Contractor's Project Manager's point of view this method has the advantage of laying the foundation for a more fact based and fair project completion time and acceptance date and the settlement of the final accounts.

5.14 Regional Differences

There are definitely different business or project cultures between continents, regions and countries. An example is the rule in Zimbabwe that, in order to enter certain restaurants or nightclubs, it is required to wear jacket and tie or traditional African outfit (incl. safari dress with long trousers). A dark blue safari suit will definitely comply. But in Nairobi, Kenya such a dark blue safari suit is considered the uniform of drivers and therefore totally unsuited for company managers contrary to Zimbabwe.

Style and formalities vary from country to country, continent to continent. Another example is punctuality. In Germany punctuality for a meeting is a must whereas in other countries a little lateness is a normal sign of importance and seniority. But the use of local traditions must be handled with great care in order to avoid being impolite.

Working habits and problem solving methods in business vary in line with regional cultures. As an example Asians seem to be closed and secretive people in their habits and their methods except between very close friends. North Europeans and North Americans seem to be more open, talkative and direct in their relation to other persons.

Regional differences in risk assessment have been studied systematically and scientifically (Yates 1993) and they found a markedly higher degree of overconfidence among a Chinese test panel than a USA test panel!

With the increasing globalization these differences might with time diminish or even disappear. But irrespective of regional aspects there are definitely significant differences between individual persons involved first of all in style, habits and methods applied and to a lesser extent in substance.

In South America the long term personal relations between the key persons are extremely important and very critical for commercial success, not only in selling and contract negotiation but also in the execution of a contract. The personal relations are a key factor in conflict prevention and in finding of compromises and settlements. This aspect is normally underestimated in the boardrooms of major USA, British and Continental European Groups.

Although the style is very different and much more impatient in Asia, the importance of long term personal relations and trust is nearly at the same level as for South America.

Consequently the interest in and study of the other party's key staff is important, when it comes to preventing or handling conflicts and making settlements. Turner (Turner 1993) describes important factors as power distance, individualism, masculinity, uncertainty avoidance of managers from different countries. Factors, which we recommend to observe on an individual level. But be careful, the high profile individualist, uncertainty avoiding Latino superman with a big taste of power, does also exist in non-Latin countries!

5.15 Case Studies

This book contains a number of project case studies of which one has already been used in Chap. 4. The case studies are prepared by one of the authors, who was directly involved in the projects when implemented years ago. Therefore they are subjective, it must be admitted, but to a lesser degree now - years after they happened. The case studies are built up in 3 sections. First the facts about the projects, its partners and the conflicts are described. Second comes a description of the conflict development and third the writers' comments and lessons learned.

The presented case studies come from projects in North Africa, in the Middle East, in Europe and in Asia. After the presentation of the project case, there will be a short discussion about regional differences in conflict prevention and settlement making.

The purpose of this section at the end of Chap. 5 about settlement negotiations is to discuss these approaches and methods in the context of real project experiences and to inspire the readers to think about the recommendations and transform them into use in their daily project work, handling disagreements and conflicts.

5.15.1 The Raw Material Plant

A) Description of the Conflict

- On a major process plant turn-key project in Africa a number of disagreements between the European main contractor and the South European civil subcontractor on invoicing the civil works and building works by bill of quantities, on scope of work, on variations, on milestones and on delays developed into a general conflict, threatening the progress and the project completion. The project was about 10 months delayed. And the civil contractor had started international arbitration on scope, delays and compensations.
- Site claims officers on both sides flooded each other's site organizations with numerous accusation type claims letters but face-to-face contacts between the two parties about the conflict or the underlying disagreements were practically non existent.
- A specific case, regarding installation of defective building materials, came up and the main contractor requested immediate replacement due to faulty workmanship or materials by the civil contractor free of charge, which the latter refused without a variation order claiming that the damages to the building materials were caused by the main contractor's operation.
- The main contractor then proposed that the arbitration tribunal should appoint an independent expert to decide the reason for the defects/breakage. The expert was appointed in agreement and his conclusion was that the main contractor's reasons were correct. As a consequence the civil contractor rectified the defective building materials installed on his own account.
- Simultaneously the parties were improving their business relations on the construction site and consequently achieved satisfactory progress.
- The main contractor was encouraged by the first expertise and then proposed that the Tribunal again appointed an independent expert to decide the reason for the leaking service tunnels underneath the plant. The expert was appointed on mutual agreement and his conclusion was that the civil contractor's reasons were right. The result of this expertise came as a great surprise to the main contractor, who believed he had prepared himself carefully including asking another

civil contractor to analyze the case with a conclusion in favor of the main contractor. Cost of rectification – kept to a minimum – was then borne by the main contractor.

- The process plant was finished and commissioned with a delay of 10 months. It was able to produce an excellent product quality, at a rate exceeding the guarantee. The client was extremely satisfied and signed a major technical assistance contract with the main contractor.
- The international arbitration on scope of work, delays and compensations between the civil contractor as claimant and the main contractor as defendant was finally decided after 8 years, and the compensation amounts were settled by direct out of court negotiations. The arbitration costs were very high for both parties.

B) The Main Elements of the Conflict

- Both parties experienced financial problems either on company level (the civil contractor was merged twice into bigger groups) or on project budget/result level.
- Apparently none of the parties sufficiently analyzed the position or interests of the key staff personalities of the other party and its consequences (ref. Chap. 4).
- None of the parties used internal Monitors of Litigation at the early stage (ref. Chap. 6) or an external Mediation regarding the main claims (ref. Chap. 7).
- The key staff on both sides did not possess or was not allowed to use a "give and take" behavior.

C) Comments on Lessons to Be Learned from the Case

- The parties could have benefited, if they had faced reality at an early stage and had recognized the upcoming conflict with the chance of taking preventive actions
- Use of a "give and take" attitude at an early stage, use of a monitor of litigation and use of better relations between the key persons of both parties could have prevented litigation and saved significant costs and management time
- An expert institution provided for in the contract could have helped the parties much more than the "ad hoc" experts used.

5.15.2 The Mineral Processing Plant

A) Description of the Conflict

- During design and construction of a mineral processing plant the Owner and the process supplier/Contractor experienced two conflict situations: first on the design, and later on the quality of the installation of the lining in a process vessel of the plant. Both conflicts were finally resolved by negotiations, but in at least one of them the parties were on the brink of "a hot litigation" during project implementation. The last situation also led to a conflict between the lining material manufacturer and the supplier.

- The contract was an Owner biased FIDIC type of turn-key contract. The Owner, being also "The Engineer", had contractually strong powers. The contract contained a general "fit for the purpose" clause and prescribed ICC arbitration, taking place in the Owner's country, and governed by the laws of a European country.
- In the first request The Engineer took the Contractor by surprise in his demand for a more expensive lining in a much greater area of the vessel than foreseen by the Contractor. A requirement that most likely was unnecessary, but the Contractor was not prepared to defend his original proposal in the right contractual way, stating that "this solution is fulfilling all the specifications" and the "fit for the purpose" requirement. Consequently the changes requested by The Engineer are unnecessary!
- In the second request The Engineer rejected the lining as installed, because the joints did not fulfill the supplier's and the designer's own specification regarding the width of the joint. A width of 2 mm was prescribed but no tolerance was specified, so areas with measurements of 1,5 mm and 2,5 mm were refused. In this case the Contractor was much better prepared and had done his homework. In order to reinforce his position he openly arranged and financed two independent experts of different nationalities to give their appraisal and beforehand declared that he would without hesitation follow their possible recommendation for necessary improvements.
- The experts generally accepted the installed lining as "fit for the purpose" but still recommended minor modifications. Consequently the contractor informed The Engineer, that he challenged his decision and that if the lining was not approved the case would be brought to arbitration. Then The Engineer backed down and accepted the lining without any modifications, not even the minor ones suggested by the experts, and agreed in principle with the Contractor.

B) *Main Elements of the Conflict*

- The Owner tried to gain time because of a general logistic problem of the plant, which was under his responsibility.
- The power game picture was blurred by a "mail box" main contractor formally acting in between the real parties (ref. Chap. 2).
- In the first situation the Contractor did not understand the correct contractual procedure and did not analyse the position of the other party and the consequences (ref. Chap. 3).
- Apparently none of the parties analysed the interests and key staff personalities sufficiently (ref. Chap. 2).
- There were limited social or personal relations between the key staff on both sides.
- In the second request the contractor's key staff was better prepared and allowed to use a "give and take" behavior (ref. Chap. 5).
- As the lining material supplied was off-specs in terms of measurement it had to be reworked on site by the Contractor to fulfill the specs of 2 mm (no tolerance). This of course made the Owner very interested in the case and gave him inspiration to raise the issue and disapprove some lining already installed.

- The contractor tried in vain to settle his claim of extra rework costs with the manufacturer of the lining. The Contractor then brought the case to court litigation where the Contractor lost. The Contractor then appealed the decision to a higher court instance and called the manufacture's Quality Assurance staff for witness hearing, which appeared to be quite damaging for his position. After that the parties agreed to negotiate and settle the claim, which was done.

C) Comments on Lessons to Be Learned from the Case

- The 2 requests described and the issues experienced are quite typical in international engineering, supply and construction contracts.
- It is recommended to analyse an upcoming conflict situation carefully, decide a strategy and execute the strategy instead of "playing it by the ear" and try the negotiation way at a chosen point.
- Keep negotiation contact the whole way through.
- Settle the disagreement as soon as possible and get on with the project progress and completion.
- Sometimes a short litigation (or part of) is necessary to get the party to the negotiating table!

5.15.3 The Food Processing Plant

A) Description of the Conflict

- In a food processing plant turn-key project in South-Western Asia a number of disagreements on alleged deficiencies of the nearly completed plant between the North European Contractor/Process Supplier and the local Owner appeared at the end of the installation.
- The Contractor found it very difficult to settle the series of disagreements because there were too many vague claims that he found were outside the scope of the contract. The owner threatened to call the 15% performance bank guaranty (on demand type) by requesting "pay or prolong" a couple of times. The contractor was forced to prolong.
- The contractor proposed a joint mediation by 2 independent experts with proven practical operational and maintenance experience in similar plants and working for a reputed international independent inspection and quality assurance company. The Owner did not react but maintained his position and again threatened to call the performance bank guarantee.
- The contractor ordered the proposed expertise by the 2 independent experts on his account and it was carried out and nearly completed (approx. 90%) when the Owner expelled the experts by applying threat of violence (local tribesmen as plant guards appeared armed with Kalashnikov submachine guns pointing at the experts). The expert's report was issued and the Contractor right away confirmed his willingness and on his own account to make the few minor modifications suggested by the experts, who substantially judged, that the plant had been delivered and installed according to the contract and was fit for the purpose.

- Then the contractor's project manager approached the owner himself and they agreed to meet on neutral territory. This leads to an agreement on some minor modifications (other than the expert's recommendations) and to the release of the bank guarantee. Later the plant was accepted by the Owner.

B) Main Elements of the Conflict

- The Owner experienced serious financial problems jeopardizing the production start (ref. Chap. 2).
- For special reasons there were only limited social or personal relations between the key staff on both sides and especially between the Owner himself and Senior Management of the Contractor (The Contractor's Execute Vice President who negotiated the contract with the Owner personally had retired and was not used for mediation despite being available).
- The key staff on the Contractor's side had conflict resolution experience and was allowed to use a "give and take" behavior (ref. Chap. 5).

C) Comments on Lessons to Be Learned from the Case

- Analyse the upcoming conflict situation carefully, decide a strategy and execute the strategy.
- Keep negotiation contact with the other party the whole way through and use external experts as witnesses.
- Settle the disagreement as soon as possible and get on with the project to reach completion (could in this case have been achieved earlier).
- Keep an open eye on the tactical situation and utilize openings for negotiations to reach a settlement.

5.15.4 The New Technology Plant

A) Description of the Conflict

- In a small turnkey project for an industrial client in the Mediterranean region the Supplier/Contractor experienced severe difficulties in achieving the guaranteed production in terms of quantity for new technology equipment installed in a process line. The gap was significant and the stop of production was at stake.
- The Owner was very upset and dissatisfied and requested immediate action by the Supplier/Contractor to remedy the defects and threatened to call the "on demand" performance guarantee.
- The Supplier/Contractor chose to work closely with the Owner to try to maintain the maximum production possible and to keep the Owner informed of his remedial actions and plans.
- The Owner's production had first priority and process and/or equipment modifications had second priority.
- The Supplier/Contractor kept a very close contact at Project Manager and Project Director level with their counterparts at the Owner's organization.

B) *Main Elements of the Conflict*

- The Supplier/Contractor used newly developed process equipment not yet fully tested and proven in industrial application.
- The Owner had not been informed that he had bought an insufficiently tested new technology.
- The contract had normal commercial conditions and did not take the testing and developing aspect into consideration.
- The Owner felt insecure about the situation and the risks but did not seriously consider to cancel the contract because of the high degree of the Supplier's/Contractor's engagement.
- The Supplier/Contractor hesitated to inform the Owner of their own uncertainty regarding the solution finally implemented and working satisfactorily.

C) *Comments on Lessons to be Learned from the Case*

- Commercial projects are definitely not suited for process and equipment development, especially not for newly designed equipment.
- The Supplier/Contractor had to bear the extra costs with no compensation from the Owner, but avoided contract cancellation or court cases and penalties.
- In other cases the authors have experienced projects where the Suppliers/Contractors have decided or had been forced to give up for financial reasons or lack of resources and have ended up in lengthy and costly litigation, costing more than the extra cost would have been if they had followed the same path as above.

5.15.5 The Semiconductor Project

A) *Description of the Conflict*

- A major US – Thai Joint Venture (JV) semiconductor project in Thailand was breaking the grounds in 1996/97.
- The US-Thai JV hired a US Main Contractor for the installation of the imported process equipment as well as for the local auxiliary equipment.
- The US Main Contractor established a 100% affiliate company in Bangkok, Thailand.
- The US Main Contractor engaged a Thai Mechanical – Electrical Engineering and Installation Company (M&E Contractor) to engineer and execute the electrical installation and the air conditioning as his subcontractor.
- When the buildings were coming to their final stage, the US Contractor ordered the Thai M&E Contractor to start to install electrical conduits and prepare the electrical installation on a small scale and on the basis of time spent and agreed daily rates before the arrival of the imported equipment.
- Work for approximately THB 7 million was carried out and invoiced by the Thai M&E Contractor, but not paid.

- Due to the financial crisis that hit Thailand in mid 1997 the work was stopped, the site abandoned, the project cancelled and the US Main Contractor withdrew from Thailand leaving the Thai M & E Contractor with an unpaid debtor position of approximately THB 7 mil. (severely needed during the financial crisis in order to survive).
- The Thai M&E Contractor presented his case in a very comprehensive case description to a top New York law firm known by the Managing Director of the Thai M&E Contractor. The law firm accepted to represent the Thai M&E Contractor for a nominal fee in the start phase.
- The strategy was to go after the US Main Contractor's main company in the USA, as US Courts normally accept jurisdiction when US companies are involved. A notification letter from the New York law firm was sent and followed up.
- After a few weeks the case was settled between the parties for approximately THB 3 million and paid up by the US Main Contractor.

B) Main Elements of the Conflict

- Start of a new cooperation between two companies from different continents and cultures.
- Although the long term contractual relationship was not yet well defined between the partners, an order to start work on preliminaries was issued to the Thai M&E Contractor.
- When the financial crisis hit Thailand, the US Main Contractor kept silent not informing his subcontractor of the possible fate of the project
- The Thai M&E Contractor's networking was not sufficient to warn about the risk of a project collapse and especially of the sudden escape of the US Main Contractor.

C) Comments on the Lessons to be Learned from the Case

- Clear and short payment conditions and a careful follow up on unpaid invoices reduce the risk significantly.
- The Lawyer's action should have been taken earlier.
- Better and safer payment conditions could have been considered but were apparently unrealistic.
- A personal relation between the expatriate staff of the Thai M&E Contractor and the Senior US Management Staff of the Main Contractor existed, but it was not sufficient.
- Doing business and especially developing new business, as the Thai M&E Contractor did, always involves the risk of doing business with the wrong people.
- The Thai M&E Contractor obtained a payment of THB 3 million in a settlement with the US Contractor's Head Office out of the approximately THB 7 mil. already invoiced. The Thai M&E Contractor did not have the management or financial resources to fight the case in a US court.
- Awareness of regional cultural differences is important, but more important is it to avoid the "business cowboys".

- In this case it is hard to say how the conflict could have been avoided, but it seems, that the Thai M&E Contractor had no other chance, than using the US Law Firm to "force" the US company to agree to negotiations.

5.16 Conclusion to Chapter 5

The conclusion of this chapter is that settlement of disagreements or conflicts by commercial negotiations and agreements requires better interest in and understanding of conflict sources, causes and stages in order for the parties to make an active intervention to stop the conflict escalation, before it destroys the cooperation of project implementation and ends up starting a litigation or arbitration.

"The strategic and pragmatic approach" is described and recommended as a general project management policy and practical approach. One of the most important practical measures to implement "The strategic and pragmatic approach" is systematic and regular project and construction site meetings with an agenda and minutes of meeting signed by both parties.

Another important prevention factor is the better understanding of the other party's position, key staff and company situation.

A commercial negotiation requires careful preparation and this includes realistic evaluation of the other party's position and claims and then of one's own. Here there is "room for much improvement". In many cases a party over evaluates its own position, and/or there is hardly any evaluation of the other party's claims, considering these claims as unjustified and rejecting them without any objective analysis. This has lead to many negative surprises later on!

The chapter contains a number of recommendations, forms and checklists on how to improve the negotiation process by organising, evaluating, building trust and applying good practises including constructive negotiation meeting agendas and minutes of meetings. Advice is given on the points to be included in the settlement agreement and attention is drawn to the importance of drafting the agreement in such a way, that it finally resolves all the issues and disagreements involved by its proper execution and thereby not giving raise for new disagreements.

The chapter also deals with handling of a negotiation break down, how to resume the negotiation later and how to conduct negotiations parallel with litigation or arbitration. This is not only allowed but also normally encouraged by judges and arbitrators.

Negotiations of delays and extension of time is a very difficult type of settlement negotiation and is therefore often postponed to the end of the project in conjunction with final accounts settlement. This is in our opinion wrong for two reasons – the first is that the positive effect on progress by a settlement will be lost, and secondly that such a postponement risks to mix up the delay issue with other issues resulting in an unfair solution for the contractor.

Finally five case studies illustrate conflict handling, negotiations and dispute resolution (by negotiations or by litigation/arbitration) supporting our general conclusion.

Settlement by negotiation is difficult and painful for both parties. It requires a lot of hard work, deep creativity and plenty of understanding of the other party's position and situation. The result seems not at all satisfactory for any of the parties, but it will normally be considered reasonable and prudent, given the circumstances; despite the emotional hardship. But most important it is productive for the project completion and quality.

No wonder why the parties during this process are looking at litigation as the best way to establish justice by a professional judge or by professional arbitrators. But in most commercial cases they do not establish justice i.e. who was right and who was wrong and their proceedings take years and are very costly. The final result in form of the verdict or an award on the project conflict claims might not be very different in form or content to a compromise the parties could and should have agreed upon much earlier by themselves. This especially applies to delay conflicts, an area of increasing importance. If it is made by the parties themselves, right away after the event, the result is strengthening the cooperation to the benefit of the project.

We recommend that the Management of the parties involved in international projects resolve their disagreements or conflicts by a negotiated agreement. This process is basically similar to the one characterizing their daily business. The earlier they take the initiative to negotiate and settle, the better. A proverb says that "the master comes from practise"!

5.17 Questions on Chapter 5

1. Identify 5 indicators, which describe an unresolved disagreement as risking to evolve into a conflict.
2. Describe 5 practical countermeasures, which can be used to prevent a conflict.
3. Write 3 (of which minimum 2 were unsuccessfully handled) case studies about conflict handling from your own professional or private life along the same steps as in the Case Studies in Sect. 5.15 above.
4. How can your future conflict handling process and commercial negotiations be improved in order to reach a settlement agreement faster and smoother? Please mention 5 initiatives.
5. Please draft a settlement agreement that you believe can be accepted by the other party in one of the unsuccessful case studies in Question 3.
6. Have you been involved in a delay claim situation involving a request for extension of time and what was your experience? Write a Case Story following the same steps as in Sect. 5.15 above.
7. How can handling of such a situation (Question 6) be improved?, base your answer on Sect. 5.13 "Negotiation of delays and extension of time".
8. Do you have experiences with local traditions and customs, influencing and interfering with commercial settlement negotiations and what was your conclusion?

References

Felding F, Rindorf E, Sejersen E (2002) Eksport- og Projektfinansiering, Published by the Authors, Copenhagen.

Elkjaer M, Felding F (1999) Applied Project Risk Management, International Project Management Journal Vol. 5, No. 1, Finland.

Felding F Project Management training course material. www.finnfelding.dk.

Yates J F, Ju-Whei L, Hiromi S, Patalano A L, Sieck W R (1993). Cross Cultural Variations in Probability Judgement Accuracy. Beyond General Knowledge Overconfidence? Organisational behaviour and human decision processes. Vol. 74, No. 2 May 1998. McGrow Hill Book Company.

Turner J R (1993) Chapter 22: International Projects. In: The Handbook of Project-based Management. The Henley Management Series.

6 Litigation, Arbitration and Mediation Contributing to Conflict Settlement

Abstract. Chapter 6 discusses the various existing procedures to solve conflicts in a major project. Whilst it is not the scope of this book to analyze the legal aspects of conflict resolution, this is left to the lawyers, it is the aim to look at conflicts from a Senior Management's and Project Manager's point of view and to give him assistance in solving the conflicts with his means or to prepare an eventually unavoidable litigation in the best way for his company. Chapter 5 has stressed the various aspects of successful negotiations for conflict settlement and this Chap. 6 will elaborate on how to prepare and to pursue successfully legal proceedings, be it in front of a state court, in front of an arbitration tribunal or in mediation.

The authors consider the role of a "Monitor of Litigation" as crucial for successful management of a litigation case. Many parties make the mistake to use the Project Manager as the key person to run the litigation, since he is considered to be the one, who knows the project best. Whilst this is true, he has a major disadvantage, he is biased and his views might be obscured by personal interests, mistakes he might have made, burdened by difficult relations to persons from the other party, etc. We shall therefore elaborate on the reasons to use a Monitor of Litigation and on his tasks in the course of litigation.

We shall then discuss aspects of a strategy, if legal proceedings seem to be unavoidable. This strategy will be decisive for which form of conflict resolution will be preferred, a soft resolution method (Sect. 6.3) or a litigation at a state court or an arbitration at one of the arbitrations institutions (Sect. 6.4). The resolution will, of course, not only depend on one's own decisions but also on the decisions of the party one is in conflict with. Section 6.5 develops two case studies and some questions to students.

Key words: Mediation, Litigation, Monitor of Litigation, Corporate Pledge, Mediator, Nomination of Mediator, ADR, Alternative dispute resolution, Legal proceedings, Arbitration, ICC, Project Manager, CIETAC, Disputes, Early Warning System, Litigation Cost, Expertise, Net Work Analysis, Third Party Inspector, Chinese Law, American Arbitration Association, International Commercial Arbitration Court, ICAC, Chinese International Economic and Trade Arbitration Commission, Corporate Pledge Model, Terms of Reference

6.1 Project Manager and Monitor of Litigation

6.1.1 Comparison of the Tasks of Project Manager and Monitor of Litigation

The Project Manager is the key person with respect to success or failure of a project. His interest is the execution of the project in the planned contractual time frame, within the budget constraints and in line with the contractual technical and quality specifications. Throughout the project the Project Manager is in constant struggle with his client and his subcontractors to achieve a successful project execution. He and his team naturally commit mistakes, as do the managers on the client's side as well as on the subcontractors' side. Such mistakes or omissions in the contract lead to disputes and might eventually lead to litigation.

From the above it is evident that the Project Manager is not in an unbiased position when it comes to conflicts between the parties. Top Management of the contractor, i.e. the superiors of the Project Manager, therefore need an "early warning system" of conflicts in order to intervene before major damage is inflicted. It is not the scope of this book to go into detail of such systems, it may only be mentioned, that a number of elements for such a warning system should be in place at the beginning of project execution. Such elements are given in Schedule 6.1.

It is Top Management's role to conclude from the early warning system, when to bring in a Monitor of Litigation (ML).

The Monitor of Litigation has to come into the scene at a very early stage, when conflicts become apparent. He or she should be a person experienced in project management and it should not be the lawyer of the company, or an outside lawyer. The monitor's role is to analyse the conflict from an unbiased position and to give advice to the Project Manager and to Top Management of how to act with respect to the conflict. During his analysis the Monitor of Litigation checks not only the company's own position with respect to legal, technical, financial, social, target date, etc. aspects but also, and very important, he checks the documentation regarding the conflict in order to have a sound written basis in case of legal proceedings with all the proof necessary to defend the case. Schedule 6.2 gives a summary of tasks to be executed by the Monitor of Litigation before legal proceedings start.

The advantage of a Monitor of litigation is not only the unbiased view of the problems, which could result in legal proceedings, but it is also his time availability for doing the tasks in Schedule 6.2 without conflicting with the day to day work of the Project Manager. If the Project Manager were charged to make the preparations for legal proceedings the result would barely be of the necessary quality and his tasks as Project Manager would suffer.

The Monitor of Litigation's unbiased view of the problems tending towards litigation is of utmost importance to give Top Management a clear view of the situation, before coming to a decision, whether to proceed towards legal proceedings or to search an amicable settlement. The Project Manager has his opinion formed by the day to day events on site, he will have his very special view of the reasons for the

Schedule 6.1. Elements of an Early Warning System

- Tight regular bi-weekly or monthly budget control (frequency depending on project size)
- Regular budget-plan, time-schedule and progress control till project end
- Detailed network-analysis and regular network update
- Systematic and well structured reporting system to the client, to the subcontractors and to the own Top Management
- Regular visits of Top Management to the construction site

Schedule 6.2. Tasks of the Monitor of Litigation

- Analyse thoroughly the problem causing the call for litigation
- Interview PM and PM's staff with reference to the problem
- Collect all necessary proof (photos, material-specimen, etc.)
- Assemble a special file for litigation with all relevant documents
- Check for complete history of the project and the litigation problem (who was involved, which documents existed before the contract was drawn up, etc.)
- Avoid disclosing his activity to the opposed party
- Collect outside documents such as e.g. weather charts, news from radio and TV, customs documents, etc., which could be of importance
- Develop a strategy for conflict solution

problems, he might try to cover up for mistakes he or his staff have made during execution, he might have a strained relation with the Project Manager or with some of the staff of the opposing party, he might have a weak documentary basis for the case, which he would want to avoid to become apparent, etc. The Monitor of Litigation has to be free of all these constraints, to which the Project Manager might be subject to. Needless to say that the Monitor of Litigation should not be someone who had been involved in the project, it should not be a former Project Manager of the project, who had been replaced, nor should it be the Sales Manager or any former staff member in the project. All this would only reduce his authority in performing his task.

6.1.2 The Monitor of Litigation

The ideal profile of a Monitor of Litigation is someone, who

- has managed projects himself,
- has sufficient technical know-how of the project in question,
- has legal experience from former litigations or other sources,
- is skilled in negotiation,
- is diplomatic in his approach,
- has sufficient free time available to dedicate to the monitoring task,
- is not too young,
- has not been involved in the planning and/or execution of the project,
- has a certain seniority in the company,
- has the full support of Top Management,
- has a wide range of decision authority and
- who is available till the end of the legal proceedings.

Naturally the Monitor of Litigation has his costs, but this cost is well spent, if he does a correct job and thus avoids the cost, which a badly prepared process in arbitration or at court can produce. (One $ spent in preparation of a process is worth ten $ of the final judgement.) Furthermore the cost of the expensive international lawyers can be considerably reduced by the Monitor's solid preparation of the case. It is often seen that the international lawyers are requested by Top Management to prepare the pleadings for court without having received a document which is prepared in-house by the company and which gives all the necessary input for the pleading writs. Considering the hourly cost of an internationally experienced lawyer of somewhere around 350 €/h, it becomes obvious that the Monitor of Litigation with a cost of around 150 €/h is a wise investment.

The Monitor of Litigation is ideally an in-house person with the above described qualities. If such person is not available, it can be an outside person as well. Persons with the above given profile can be found among the many independent consultants from small consultant bureaus or "one-man" shows. Such an outside Monitor of Litigation can perform all the tasks listed in Schedule 6.2 and thus in-house manpower does not need to be withdrawn from their regular tasks.

With respect to the Monitor of Litigation it should finally be mentioned that he should have experience in court proceedings and arbitrations besides his experience as project manager. This experience will facilitate his task of checking the documents and the evidence and possibly making the necessary corrections, adjustments or additions. He is also the person to evaluate the pleadings of the legal adviser, once these have been written.

6.2 Considerations Concerning the Strategy for Litigation

When in conflict, people tend to consult a lawyer immediately without having well analysed their case themselves. The lawyer will then give his opinion on the case and in many cases suggest to initiate legal proceedings. Not all lawyers will give a completely unbiased opinion. To really be able to analyse the chances in litigation one should first build up one's own strategy, on the basis of which the discussion can be held with the lawyer.

It should be kept in mind, that it is not the lawyer who wins or loses the case, but it is the suing party. The lawyer assists with his legal knowledge and process experience, but the evidence that makes a case a winning case comes generally from the party involved. The authors believe strongly that the strategy has to be developed by the party and then discussed with the lawyer, since the party knows best what has happened on site, and not the other way around.

The strategy should be developed by the Monitor of Litigation in cooperation with the in-house legal adviser, if the party has such an adviser, if not, the strategy should first be developed before consulting the external lawyer. Relevant points to be considered in developing the strategy for litigation are compiled in Schedule 6.3.

The different aspects to be considered when developing a litigation strategy as listed in Schedule 6.3 are self-explanatory so we do not discuss them in detail here. Nevertheless some observations and recommendations are made.

When describing the case at stake, note that specific attention should be paid to an unbiased check whether the proof available is a real proof for the case and whether the other party has more evidence to prove the contrary. Long years' experience has shown that the party which has its evidence well documented is normally the one that wins the case. In short this can be described by "Keep documenting your evidence and prevail". This advice goes towards the Project Manager during execution, who is

Schedule 6.3. Relevant aspects to be considered in a litigationstrategy

- Description of the case at stake, including the problems occurred, explaining technically or financially why the problem is being disputed
- Listing of the claims or potential claims raised by the other party
- Listing of the own party's claims
- Analysis of the chances for the respective party in each of the raised claims, using as basis the documented facts, the contract and the laws governing the contract
- Simulation of legal proceedings requested by the other party
- Decision of, whether an outside independent expert should be consulted for advice on technical questions and documentation of physical facts
- Estimation of the duration of the proceedings
- Estimation of cost: for the claims of the other party, of the tribunal, of the lawyer of the other party, and of the own lawyer
- Estimation of receivables for the own claims
- Estimation of in-house costs for the Monitor of Litigation, for other personnel (technical, administrative, commercial, Top Management) including travelling costs, records of the own costs during the period of litigation should be kept at all means
- Estimation of implications on the ongoing project execution
- Analysis of other implications, such as image of the company, future relations with the other party (client, subcontractors), tied up personnel in litigation not available for other projects, etc.
- Estimation of whether the opponent will be able to pay in case that the outcome of the case will be positive
- Balance of potential receivables and costs
- Decision of whether an offer should be made to the other party to settle the case

bound to pay special attention to the documentation of all events during engineering and on site, even though the subject might not seem of importance.

The listing of the claims the other party could raise is of course a supposition, but a good project manager knows the problems the client or subcontractor might raise. It is very unusual that the other party does not have counterclaims in preparation as well. It is absolutely necessary to analyse, whether there could be any counterclaims before starting any proceedings.

To well analyse the possible counterclaims, it is useful to go through a simulation of a litigation, which could be requested by the other party. In this simulation one should execute the following steps (compare also to Schedule 6.2):

- Which claims could the other party raise?
- How would they prove their case?
- Which evidence do they have (documents, witnesses, photos)?
- Which contractual clauses apply?
- Has there been any acceptance of the claims by the own staff?
- Eventual consultation of an outside expert on the grounds for the other party's claim

- Consultation of the in-house lawyer on the grounds of the other party's claim
- Analysis of the claims in money if the other side were granted these claims

In some cases it might be necessary to call in an outside expert. Such expert can give an opinion on technical, financial or legal matters that might be under dispute. Most of all, though, the outside expert can serve as an expert-witness and record facts that might be covered up later on by the advancing project. Although the report of an expert-witness will be considered a partisan-expertise by a judge, such witnessing sometimes can be the only way of documentation by a third party. As experience has shown, such an expert-witness has a much higher credibility in front of a judge than an employee of the party. Naturally the best way for documentation by an outside-expert would be if the opponent would agree on the selection of such an expert. But such agreement is only a rather rare case. If the expert were accepted by both parties before starting his mission, then the judge will certainly accept the report as a true proof. By all means, the party that has engaged the expert-witness, always has the right to call upon the expert as witness in front of the court.

Whilst it is difficult to estimate, how much (especially in-house cost) the other party might request, the approximate costs of the court and the lawyers can be estimated. One should never estimate these costs on the low side, as experience has shown adding some 50% to the first estimate would be wise. The cost of a state-court can normally be inquired at the court; the cost is calculated normally as a percentage of the total sum under dispute (claim of both sides together, and counterclaims added). If arbitration is chosen, the cost of the tribunal can be calculated according to the rules of the Chamber governing the arbitration (e.g. Rules of the International Chamber of Commerce, Paris, see Tables 6.1 and 6.2). If the court might call for a technical expert, the costs for such expert (s) have to be added to the court's cost.

Estimating one's own costs at first seems to be easier. But experience has shown that the in-house costs are always underestimated. The time spent by the different in-house departments are constantly underestimated. Top Management, the design-department, the execution groups, the project management on site, the financial departments, the legal department etc. will be involved and cause costs. Travelling costs for appearing in front of court, research on site etc. are often not being considered sufficiently. Furthermore the different divisions of a company tend to use a cost centre such as the litigation cost centre as a convenient place to charge cost and reduce the cost of their own centre.

The recording of in-house costs for litigation is of extreme importance. It is not only necessary for Top Management as a decision basis, whether to go on with litigation or to stop it and search an agreement, but it is also necessary for claiming in case that the court decision is favorable. Not all the in-house cost will be accepted by the court, even if the case is won by 100%, but it is important to be able to claim them and let the court cut out, what they feel unjustified. Finally such records are valid for future litigations as a reference.

Experience shows that most of the litigation on projects, especially international ones, are started during execution. This might not be avoidable, since the cause for the dispute might have an impact on the correct execution of the portion of the project still remaining. If, for example, a certain part of the necessary construction site of the project is not available, or a local subcontractor enforced by the client does not and can not perform in the way foreseen, or a specific equipment offered in the project is no longer available, etc., then a solution has to be found before the project continues.

It is obvious that the start of litigation has an important impact on the willingness to cooperate between the litigating parties. The readiness for compromise is normally heavily reduced, the normal "give and take" necessary in any project will barely exist, since in every compromise the parties think of the legal proceedings. The very human behaviour of "trying for revenge" will come forward and poison all relations. Furthermore, during the ongoing execution, each party will try to find proof for their case under dispute from the actions undertaken by the other party. Although most legal proceedings start during the course of the project, it is recommended to avoid them during execution by all possible means.

The analysis of implications outside the specific project must be made carefully, such implications are e.g. image in the market, relation with the client or subcontractor in the future, the influence on other ongoing projects with the same client or the same subcontractor, the acceptance in joint-ventures with competitors for other projects, etc. The loss of market position can be more costly than the benefits out of the litigation. Another aspect to be taken into consideration is the binding of personnel or equipment in a project under litigation, which will not be finished in the projected time-span, thus binding the assets.

When making the decision, whether to go for legal proceedings or not, the unquantifiable costs have to be included in the balance of benefits and costs, and should even be quantified by some kind of estimation. Unnecessary to say, that the balance between potential benefits and costs should be very favourable to the benefits, a factor of 10 between benefits and costs seems to be reasonable, otherwise litigation should be avoided.

Finally the strategy for legal proceedings should carefully consider the factor "time". The duration of legal proceedings, be it in front of a state-court or in front of an arbitration tribunal, tend to be very lengthy. An argument often used in favour of arbitration is the shorter duration of proceedings in arbitration compared to state-courts. Whilst this might be true in some cases, arbitration can still be very lengthy in others. The authors have known arbitration tribunals which have lasted eight, ten or more years. Thus the time arbitration might take, should not be underestimated. In Sect. 6.4 further explication will be given on arbitration.

A rule of thumb regarding the length of arbitration and litigation is that a litigation in front of a state court will be shorter than arbitration, if the case is resolved in one instance. If there are more instances then arbitration might be shorter, since there is only one instance in arbitration.

After having well analysed all relevant aspects concerning the legal proceedings, top management will have to take a preliminary decision whether to go to court. Once they have taken such preliminary decision on the basis of the in-house analysis, only then should the outside lawyer be involved and give his opinion on the basis of the analysis presented.

Nevertheless the authors have in some very complicated cases (commercially & legally) experienced external lawyers to be useful in structuring the case from a legal point of view at an early stage.

According to experience of the authors, the questions most often at stake in litigation are to a much lesser extent legal questions than technical or economic questions, where the two sides interpret their contract differently. Only very seldom we have seen that pure legal proceedings had to decide on the non-compliance with the governing laws. In view of the fact that the claims at stake are economic or technical questions, the authors strongly recommend that solutions outside of formal litigation are sought. Intelligent negotiation or soft resolution methods are the better way than going through formal legal proceedings (state court or arbitration).

Having performed the above analysis, the final strategy will come out as a result. After consultation with the outside lawyer, Top Management will decide whether to still go ahead with litigation or not. If such decision is to go ahead, then the strategy for the litigation will be based on the results of the above given analysis. This strategy then has to be developed together with the outside lawyer who will prepare the documents to be presented to the court in cooperation with the Monitor of Litigation.

The strategy will define which of the potential claims should be pursued. Only those should be included in the legal proceedings which have a good documentary basis and therefore a chance of being won, the doubtful claims should be left out, since they will only increase the cost of the proceedings by increasing the sum under dispute (see cost tables of the ICC in Tables 6.1 and 6.2).

Next to having decided on the claims, a decision is necessary on how to approach the other party. Presumably the direct negotiations have not been successful, therefore it has to be decided whether the foreseen contractual procedure (state court or arbitration) will be selected or an intermediate step will be taken, which could be a "Mediation" or a "Dispute Review Board". These procedures will be discussed in the following Sect. 6.3 "Soft Resolution Methods", whereas the "hard" procedures will be discussed in Sect. 6.4 "Arbitration and Litigation".

6.3 Pre-Arbitral and Soft Resolution Methods

6.3.1 Mediation and Referees Stipulated in the Contract

In view of the cost, difficulties and duration of state court or arbitration procedures other approaches were looked for, which could reduce these negative effects. The so called soft resolution methods are those methods which use one or three mediators for conflict resolution, but their verdict is generally not binding for the parties.

The ICC (International Chamber of Commerce) in Paris, France offers rules for "optional conciliation", which provide for the appointment of a conciliator who can suggest terms of settlement to the parties who in turn may accept or reject them. The rules regulate the procedure the mediator or conciliator should adopt.

Furthermore the ICC offers pre-arbitral rules which are called "Pre-Arbitral Referee Procedure". These rules are designed for the parties to have recourse at a very short notice to a third person – the "Referee" – who is empowered to order provisional measures needed as a matter of urgency. It should be noted that this procedure is binding until a subsequent ruling of a court or arbitral tribunal has decided otherwise. The decisions of the "Referee" can of course be non-binding if the parties should so decide in their contract. But these rules are directed mainly to the urgent cases which cannot wait for any extensive analysis such as in a court or arbitral tribunal.

The clause to be put into the contract for "Pre-Arbitral Referee Procedure" of the ICC reads:

"*Any party to this contract shall have the right to have recourse to and shall be bound by the pre-arbitral referee procedure of the International Chamber of Commerce in accordance with its Rules for a Pre-Arbitral Referee Procedure.*" (ICC)

Whereas the pre-arbitral procedure of the ICC is directed towards the urgency of disputes, several other organisations have been active in promoting the soft resolution methods. The well known American DRB Foundation, Seattle, Washington, USA is such an organisation which helps in providing procedures, contract text for including the procedure in the contract and who can provide for individuals who could serve as mediators. DRB stands for "Dispute Review Board". The approach of the DRB Foundation in conflict solving is somewhat different to the ICC approach. Whereas in the ICC rules the referees, mediators, consultants, etc. intervene only if conflicts arise, the DRB follows continuously and regularly the implementation of the project (by way of site visits, monthly reports, correspondence, etc.). As a general rule DRB recommends visits on international projects every 3–4 months at minimum. In this respect the DRB acts similarly, but not as intensively as a "Third Party Inspector".

In some cases the mediator(s) is (are) already mentioned in the contract, whereas in other cases the mediator(s) will only be nominated in case of need. The mediators can be lawyers, engineers and/or economists, the important issue is that these persons must be experienced with major international/national projects to understand the problems and requirements of such projects.

It is recommended that the mediators be nominated in the contract. Nomination at the moment of conflicts tends to "bring sand into the gear". There is generally one party which is less interested in solving conflicts than the other. This party then will try to delay the nomination of the mediators, thus gaining advantages on their issues.

The mediators are referred to by a party for a decision on a certain conflict, which has come up in the project. In general they want to have a decision soon. If the mediator first needs to get acquainted with the project, when he had not been nominated in the contract, then the time needed for mediation will be considerably longer than in the case of contractual nomination. A considerably better approach with respect to time is the DRB approach of a regular project-surveillance of the mediator. Whilst this approach is faster, it is also much more costly, since the mediator or mediators have to be paid even if no conflicts arise.

Special attention should be given, when a mediation clause is included in a contract, which will be executed in China. As will be explained more in detail in Sect. 6.4 Chinese Law requires, that the arbitration commission has to be designated already in the contract. The authors therefore strongly recommend that in case a mediation clause is foreseen in the contract, the referee (s) should by all means be designated in the contract as well.

Experience shows that successful mediation is performed in two steps: First the mediator makes a proposal in principle, leaving it to the parties to find a solution on the basis of these principles. (The parties are often better placed to define the details than the mediator.) If the parties do not reach an agreement, they then will refer back to the mediator in a second step to ask for a final proposal. The mediator will then request the parties to detail their demands on the basis of the before stipulated principles, which shortens the involvement of the mediator considerably since he only needs to evaluate which of the demands better meets the before stated principles.

It would go beyond the scope of this book to go into details of all the existing institutions and private organisations which offer rules and experts for the pre-arbitral or soft resolution of conflicts. It should be mentioned, though, that in many countries mediation or pre-arbitral solutions have become part of the legal proceedings of the countries. With the courts in many of the countries completely congested, the business community needs a faster way of solving their disputes. Records show that in more than half of the cases put before a mediator, depending on the various countries and their legal system, the solution suggested by him was adopted by the parties without continuing legal proceedings.

Even if unsuccessful, the cost spent on mediation normally is not lost. If the conflict continues before the courts or an arbitral tribunal, the work performed by the mediator should be submitted to the court or the arbitral tribunal and the mediator(s) may serve as a witness providing good evidence of what happened on site, when mediation took place. This way often the nomination of an expert (see Chap. 7) can be avoided, thus saving the considerable cost an expert might cause.

6.3.2 The Corporate Pledge Model

Whilst mediation or other ADR-procedures have found their way into international dispute resolution, the state courts still remain as the main resolution place for national disputes. This is especially true for Europe. In the United States, though, mediation has made much headway in the resolution of commercial disputes. In fact, some authors describe the United States as the "motherland" of mediation [2].

Mediation has found so much acceptance in the business community of the United States that many companies have made it their standard resolution tool. There mediation is used for resolving problems between companies of different ownership as also for companies of the same ownership, i.e. companies belonging to the same group. It is not the scope of this book to describe resolution methods of in-house conflicts, but the corporate pledge model which will be described hereafter is used in the United States also to resolve in-house conflicts.

The voluntary pledge of US companies to resolve disputes with other companies by alternative resolution means has been declared formally by more than 4 000 US corporations, among them such internationally active companies as BASF Corporation, Black & Decker, The Boeing Company, Ernst & Young LLP, Volkswagen of America Inc., Texaco Inc. and Akzo Nobel.[3] An institution which promotes such corporate pledges is the "CPR International Institute for Conflict Prevention and Resolution", New York, USA. CPR Institute is a membership-based non-profit organization that promotes excellence and innovation in public and private dispute resolution, serving as a primary multinational resource for avoidance, management, and resolution of business-related disputes [3]. The Institute suggests as clause for the pledge:

CPR Clause for Corporate Pledge [3]

"In the event of a business dispute between our company and another company which has made or will then make a similar statement, we are prepared to explore with that other party resolution of the dispute through negotiation or ADR techniques before pursuing full-scale litigation. If either party believes that the dispute is not suitable for ADR techniques, or if such techniques do not produce results satisfactory to the disputants, either party may proceed with litigation."

Having seen the fear of US and non-US companies to be pursued by the judgment of a US-court, the Corporate pledge model has made considerable headway in the United States. Clauses like the one given above are normally used, sometimes with slight modifications of the wording. The clause of course, is not binding for neither party but can open the way for negotiations under assistance of a referee, which in many cases has helped to resolve the conflict and thus to maintain a business relation, which might have broken, if a formal litigation had been undertaken.

6.4 Arbitration and Litigation

6.4.1 General Considerations

Whilst the parties have a predominant role in a pre-arbitral or mediation procedure, their influence on the procedure at court or in front of an arbitral tribunal is much reduced. The parties become subjects of the court.

In this section about court or arbitral proceedings the authors do not intend to comment on legal matters. These are handled professionally by the numerous articles on cases, law interpretation, procedures, etc. published by the legal community. This book in contrast will try to give advice to the project people and the parties how to act in front of a court in order to improve their case or not to worsen it. This advice results from the many sessions the authors have assisted in as expert or as member of a party on trial.

From the explanations given above in Sect. 6.2 and 6.3 it becomes evident, that litigation in front of a court or an arbitral tribunal should be avoided as much as possible. They are costly, time consuming and don't necessarily lead to a more satisfactory result than a compromise. As the saying goes: "A bad compromise is better than a favourable judgement." But since both parties normally think that they are within all their rights, they prefer to go to court or arbitration, and only find out later that a compromise would have been better, most of all it would have been faster.

Each contract needs agreements about the way to proceed in case of conflicts. If both parties reside in the same country and execution of the project is also in the same country, then the parties might decide to resolve conflicts via the courts of that country. If the procedures of the courts are not wanted, arbitration can be agreed upon by the parties in many countries as well. Especially in the USA, Canada and the European Union private arbitration, avoiding the state courts, has become popular in the business community.

6.4.2 Arbitration

For international projects, where client and contractor are not from the same country and execution is mostly in the client's country, arbitration on the basis of the rules of one of the well known Chambers of Commerce has become the predominant way of excluding the courts of the client or the contractor, taking thus the Chamber of Commerce or any other neutral body as the neutral entity acceptable to both parties. The International Chamber of Commerce in Paris, the Chambers of Commerce in Stockholm, Vienna, Geneva (lately combined with other Swiss Chambers of Commerce in their arbitration rules), the London Court of International Arbitration, The Commercial Arbitration Court in Russia, The China International Economic and Trade Arbitration Commission etc. are such institutions who have established rules for international courts of arbitration which can be used for resolving conflicts.

As examples we give the standard clause of the ICC in Paris, France, of the American Arbitration Association, New York, USA, of the Commercial Arbitration Court in Russia and of the China International Economic and Trade Arbitration Commission, which could be included in the contract between the parties.

The ICC in Paris proposes the following contractual clause:

Arbitration Clause of the ICC in Paris, France

"All disputes arising in connection with the present contract shall be finally settled under the Rules of Conciliation and Arbitration of the International Chamber of Commerce (ICC) by one or more arbitrators appointed in accordance with the said rules."

The American Arbitration Association of the US proposes the following contractual clause for arbitration:

STD 1 Arbitration Clause of the American Arbitration Association in New York, USA [4]

"Any controversy or claim arising out of or relating to this contract, or the breach thereof, shall be settled by arbitration administered by the American Arbitration Association in accordance with its Commercial [or other] Arbitration Rules [including the Optional Rules for Emergency Measures of Protection], and judgment on the award rendered by the arbitrator(s) may be entered in any court having jurisdiction thereof."

This arbitration clause is the standard clause proposed by the American Arbitration Association, special clauses for international contracts, for contracts already under execution, etc. can be found in the cited web-site of the American Arbitration Association. The reader should note that the ICC foresees a final settlement, whereas the AAA only suggests a settlement. It is a legal question, which of the two are more binding, the authors suggest that the word "final" be included in the clause of the AAA as well.

Other arbitration institutes in the US, such as the CPR International Institute for Conflict Prevention and Resolution [3], offer their services in the case of arbitration. The clauses they use are similar, and we abstain from discussing all these clauses in the context of this book. When drawing up the contract, the contractor should refer to his legal advisor and discuss, which of the various US institutions is the most appropriate for his case. This of course involves his client, who has to be in agreement with the arbitration institution chosen.

Since Russia is a signatory of the 1958 – Convention on the Recognition and Enforcement of Foreign Arbitral Awards – the "New York Convention", arbitration agreements in an international contract with a Russian partner may be based on the ICC or other European or US American institutions. Nevertheless the Russian partner might insist on a Russian arbitration by any of the local Russian Chambers of Commerce or the International Commercial Arbitration Court under the auspices of the Chamber of Industry and Commerce of the Russian Federation (ICAC) in Moscow, the major international commercial arbitration institution in Russia. It is the successor to the Foreign Trade Arbitration Commission (VTAK) which was established in 1934.

The International Commercial Arbitration Court (ICAC) in Moscow, Russia proposes the following contractual clause for arbitration [5]:

Arbitration Clause of the International Commercial Arbitration Court (ICAC) in Moscow, Russia [5]

"Any dispute, controversy or claim which may arise out of or in connection with the present contract (agreement), or the execution, breach, termination or invalidity thereof, shall be settled by the International Commercial Arbitration Court at the Chamber of Commerce and Industry Of the Russian Federation in accordance with its Rules."

The clause of the Russian ICAC is very similar to the ICC-clause, except for the missing word "final" for the settlement.

Finally we look at the arbitration clause proposed by the Chinese International Economic and Trade Arbitration Commission (CIETAC) [6]:

Arbitration Clause of the Chinese International Economic and Trade Arbitration Commission (CIETAC) [6]

"Any dispute arising from or in connection with this Contract shall be submitted to the China International Economic and Trade Arbitration Commission for arbitration which shall be conducted in accordance with the Commission's arbitration rules in effect at the time of applying for arbitration. The arbitral award is final and binding for both parties."

We strongly recommend that the contractor and the client, together with their legal adviser, carefully study the rules of the arbitration institution they choose. Whilst the arbitration clauses, as shown above for the various arbitral institutions are very similar, the rules to be applied might be different. In the case of the Chinese CIETAC a difference comes up via the Chinese Arbitration Law. In Art. 16 this law defines as prerequisite for the effectiveness of an arbitration agreement:

"An arbitration agreement shall contain the following:

1. *The expression for application of arbitration.*
2. *Matters for arbitration.*
3. *The arbitration commission chosen"*

Whilst prerequisite 1. and 2. are common in international arbitration clauses, the third is a particularity in Chinese law. The parties have to nominate the arbitrators at the time of the contract or at least do such before starting arbitral proceedings [6].

The parties have to decide not only on the arbitral institution but also on the law governing the arbitration. It might well be that arbitration be governed by the rules of ICC in Paris but the law of substance to be applied is the German, Swiss, French or other law. Further to the law governing the arbitration, the language and place of arbitration as well as the number of arbitrators have to be stipulated in the contract. Normally each party will nominate one arbitrator and the third arbitrator will be nominated by the Court of Arbitration (ICC's committee supervising individual Arbitration Tribunals) or by mutual agreement between the two arbitrators nominated by the parties. If there is only one arbitrator, he or she will be nominated either by both parties in mutual agreement or by the Court. In all cases each arbitrator must be confirmed by the Court. There are some arbitral institutions who reserve the right of nominating the arbitrator, if there is only one, or the President of the Arbitration, where there are three arbitrators, e.g. the ICAC of Russia [5].

An item which is often forgotten in contracts is the stipulation of a procedural law applied to the arbitration. The procedural law is of significant importance though, since it might change the course of a trial, admitting some evidence and rejecting others. The procedural law can also be of a different country than the law governing the contract, though this is not recommended by the authors.

The procedural law will define e.g. procedures of the court, procedures of questioning witnesses, deadlines and extension of time to be observed by the parties, rules for the drafting of minutes, rules for drafting the court's decision, the exclusion of arbitrators, representation of a party in front of the court, the use of experts in front of the court, powers of attorney, regulations regarding the cost distribution between the parties etc. It should not be underestimated that the procedural law to be applied to a trial has great importance, since the procedures might sometimes decide a trial in favour or against a party. The parties need advice from international lawyers with extensive experience before choosing the procedural law.

Before proceeding to the preparation of the case the arbitrators will draw up, on the basis of the documents and/or in the presence of the parties, a document defining their Terms of Reference. This document is of utmost importance to the parties since it names the respective claims of the party and the issues to be determined. The Terms of Reference will be signed by the parties and the arbitrator(s). The parties have to absolutely verify whether the Terms of Reference correctly define their claims. In general no claim beyond the Terms of Reference may later be submitted to the same Tribunal.

Many of the international disputes, at court or in arbitration, involve technical, economic, financial, legal, medical, environmental, etc. questions to be resolved which go beyond the knowledge and experience of the arbitrator(s). In these cases the tribunal or court will appoint one or more experts to resolve these special matters and report to the tribunal. Many cases have been decided to a great extent by the expert. Therefore again, it is of extreme importance that the parties check thoroughly the qualification and suitability of the expert. We refer to Chap. 7 of this book where the question of expertise is treated in more detail.

6.4.3 Litigation

As is common practice we speak of "litigation", when we refer to trials at a state court of a country. Since such trials at state courts reduce in importance in international contracts and the commercial laws differ considerably from country to country, we will not go into detail of these laws, this would go far beyond the scope of this book. The legal advisor is the person to give advice to Project Management and to the Monitor of Litigation, when litigation is undertaken.

A typical clause regarding the governing laws and regulations with respect to possible litigation could be:

Legal Clause for Application of a Country's Law

"The validity and legal interpretation of this Contract will be governed by the laws of... (country to be named)."

Many of the suggestions we give here, of how to handle legal proceedings from a Project Management's point of view are applicable as much for arbitration as for litigation. Therefore we shall now look at some recommendations we want to give to the parties with respect to legal proceedings in major projects. These recommendations stem from several years of experience in arbitration of major projects.

6.4.4 The Choice of Arbitrators

By observation the authors have noted that the parties have frequently expressed their discontent with the judges or the arbitrators. Whilst they have no influence on the choice of the judge of a state court (unless they can reject a judge for being biased), they have, in the case of arbitration, influence on the choice of at least one arbitrator, if three arbitrators form the tribunal. Therefore a party should always insist on having three arbitrators form the tribunal, unless the amount on trial is really low, e.g. below 100 000 €.

What were the reasons of discontent with the arbitrators? The most frequently mentioned criticisms were:

- lack of time of the arbitrators, thus prolonging the arbitration unnecessarily
- insufficient knowledge of the subjects to be decided, leading to problems in the Terms of Reference or in the final decision (award)

- insufficient knowledge of the contract law or the procedural law, leading to misinterpretations of the law
- excessive age of some arbitrators in some cases
- insufficient knowledge of the contract language in international projects

The party should be aware of the capabilities of the arbitrator they propose, they should be sure that he is willing to accept the task and they should verify that he is capable of negotiating with his co-arbitrators.

Needless to say that the arbitrator should be fluent in the language of the contract. Many arbitrations have been seen where one or more of the arbitrators had difficulties with the contract language. Considering the language capabilities, the examination of witnesses should also be taken into account. Important witnesses might only be capable to express themselves in their native language, which makes the examination more difficult. If translation is needed for the examination, then again the choice of a suitable interpreter is of importance. The minutes of such examination dictated normally by the President of the arbitral tribunal, have to be well understood by the witness, eventually through correct translation, so that he can confirm his testimony.

Judges and arbitrators have to be independent in their judgement; therefore they have to be free of any relation to the parties. In the law of many countries a lawyer who has worked for a party in former times can not be judge or arbitrator in a case involving this party, even if his engagement with the party has stopped before the case starts. In case a proposed arbitrator should have a doubt about his independence, he should better reject his nomination before putting into question the whole arbitral tribunal. The President of the arbitral tribunal should always be a lawyer, whereas the other two arbitrators might be of another profession. Experience shows though, that the arbitrators in general are lawyers. Generally the two arbitrators nominated by the parties will nominate the President of the arbitration tribunal. If no agreement is reached then the governing body of the arbitration (ICC, etc.) will nominate the President.

In the rules of the arbitration institution chosen, the question of partiality of the judges is normally treated at length. We recommend that the parties study these rules and regulations at the time of choosing the arbitration institution and thus to make sure that the judge they want to propose in case of arbitration can not be rejected because of partiality or any other reason.

6.4.5 The Cost of Litigation

For state-courts there is in general a cost schedule for the cost of the tribunal and in some countries (e.g. Germany) there is also a cost schedule for the lawyers. In some other countries state-courts are even free of charge (e.g. Spain). So there is no need in this publication to go into the question of costs of a litigation, since they can be looked up in these schedules. The cost for one party of course depends on the laws of the respective country. In some countries the lawyers of both parties are to be paid by the loosing party (e.g. Germany) whereas in other countries each party has to pay its own lawyer, independent of the result of the litigation (e.g. USA, France).

Table 6.1. Administrative Expenses of ICC Arbitration [8]

Sum in dispute (in US Dollars)	Administrative expenses
up to 50 000	$ 2 500
from 50 001 to 100 000	4.30%
from 100 001 to 200 000	2.30%
from 200 001 to 500 000	1.90%
from 500 001 to 1 000 000	1.37%
from 1 000 001 to 2 000 000	0.86%
from 2 000 001 to 5 000 000	0.41%
from 5 000 001 to 10 000 000	0.22%
from 10 000 001 to 30 000 000	0.09%
from 30 000 001 to 50 000 000	0.08%
from 50 000 001 to 80 000 000	0.01%
over 80 000 000	$ 88 800

In the case of arbitration the governing body of arbitration (e.g. ICC Paris) has a cost schedule for the cost of the court and the administrative fee of the governing body. These costs are in relation to the total amount at dispute. The amounts charged by the ICC can be looked up and calculated at the web site www.iccwbo.org or www.iccarbitration.org on the internet, which we reproduce here as Tables 6.1 and 6.2, the values given have been effective since January 2008. To calculate the administrative expenses and the arbitrator's fees, the amounts calculated for each successive slice of the sum in dispute must be added together, except that where the sum in dispute is over US$ 80 million, a flat amount of US$ 88 800 shall constitute the entirety of the administrative expenses [8].

Table 6.2. Fees for one arbitrator in case of ICC Arbitration [8]

Sum in dispute (in US Dollars)	Fees	
	minimum	maximum
up to 50 000	$ 2 500	17.00%
from 50 001 to 100 000	2.50%	12.80%
from 100 001 to 200 000	1.35%	7.25%
from 200 001 to 500 000	1.29%	6.45%
from 500 001 to 1 000 000	0.90%	3.80%
from 1 000 001 to 2 000 000	0.65%	3.40%
from 2 000 001 to 5 000 000	0.35%	1.30%
from 5 000 001 to 10 000 000	0.12%	0.85%
from 10 000 001 to 30 000 000	0.06%	0.225%
from 30 000 001 to 50 000 000	0.056%	0.215%
from 50 000 001 to 80 000 000	0.031%	0.152%
from 80 000 001 to 100 000 000	0.02%	0.112%
over 100 000 000	0.01%	0.056%

Table 6.2 reproduces the fees to be paid to one arbitrator. If three arbitrators are appointed then the triple of the amounts of Table 6.2 has to be taken into account. Since the president of an arbitration panel has normally a higher work load, the arbitrators might agree to have a higher amount paid to the President, which of course reduces the amount paid to the other two arbitrators.

The total costs consist of the administrative fee plus a fee for each of the arbitrators. If for example the amount at dispute is 1 Million US$ then the total expenses for the arbitration court would be at minimum 59 910,– US$ (~6% of the sum in dispute) and at maximum 201 000,– US$ (~20% of the sum in dispute), when assuming a tribunal of three arbitrators. In the case of only 1 arbitrator the cost would be between 32 970,– US$ (3,3%) and 80 000,– US$ (8,0%). The amounts for arbitration with three arbitrators are of course higher than with one arbitrator. Nevertheless we recommend to always use three arbitrators, unless the amount at stake is below 100 000,– US$. Expenses of the arbitrators (such as travelling, accommodation, food allowance etc.) and taxes have to be added to the above calculated cost of the arbitrators.

Needless to say that the sum in dispute is the sum of one's own claim plus the counter-claim of the other party, which rapidly amounts to considerable sums in international projects.

In big international arbitration tribunals it is very usual that a technical analysis by an outside and neutral expert of the problems under dispute becomes necessary. The cost for such an expert has also to be added to the cost of the ICC, since the expert is usually nominated and paid by the ICC directly. The experts charge their work on an hourly rate which is somewhere in the range of 200 to 300 €/h plus expenses such as travelling costs, telephone, printing etc.

Experts can be requested directly from the ICC or from the national Chambers of Commerce which keep files on competent experts for a wide range of disciplines.

The lawyers are in most cases paid by the party that engages them. It can naturally be agreed between the parties in their contract that even the cost of the lawyers are split in accordance to the verdict, though this regulation is quite unusual. The costs of internationally experienced lawyers are quite substantial. Typically the lawyers charge on an hourly rate, if one is lucky one might agree on a lump-sum, though this is rather unusual. The rate for an internationally working lawyer can range anywhere between 250 and 400 €/h, some top lawyers might even charge more than 400 €/h. These costs exclude of course taxes like value added tax and expenses such as travelling cost, telephone, etc.

A cost category that is always underestimated by the parties are the in-house cost. These cost include the departmental costs of the legal department, the monitor of litigation, the concerned technical departments (design department, construction department, logistic department, documentation department etc.) and the Top Management. Experience shows that the in-house costs range from three to four times the external costs of litigation. Needless to say that the mentioned in-house costs are those which can be monitored. They do not include the impact on other projects of the company, where the personnel cannot perform the way they should because they are often drawn away from their regular tasks in order to help in the preparation and the follow-up of a litigation case.

Schedule 6.4. Cost Breakdown of a major arbitration claim

A. Amount under dispute:	original claim	~70 Mill. €
	counter claim	~30 Mill. €
	total amount	~100 Mill. €
B. ICC Cost:	Admin. Cost	~0.1 Mill. €
	Arbitrators	~0.85 Mill. €
	Expert	~0.10 Mill. €
	total ICC cost	~1.05 Mill. €
C. Lawyer & other (only one party)		0.4 Mill. €
D. In-house Cost	All departments (only one party)	
	(7 Man-Years)	~2.0 Mill. €
E. Duration of Claim		~10 years

Schedule 6.4 gives a typical cost break down for an international project which was executed in an Arabic country by European companies. The plant to be erected was a raw material conversion plant.

Total Costs of the case have not been given in Schedule 6.4 because it would be misleading, the split of cost between the parties being different in each case. In case that the costs of the ICC are split half and half between the parties, then the party under consideration would have to pay 3.49 Mill. € for the arbitration. In case the party would have lost 100% of the case, then their cost would have been considerably higher, since the loosing party might have to pay the lawyer and the in house cost of the other side as well.

The example case shows that the cost of an arbitration are very considerable and should be seriously weighed against the chances of winning.

Most disappointing is the length of the arbitration in this case. This duration of 10 years is not normal, but durations of 4–5 years seem to be a normal duration of an arbitration case on international big projects. Since arbitration in most cases does not allow an appeal (unless there are formal faults in the judgement), the case is solved at the end of arbitration. Here lies an advantage of arbitration. In state litigations there can be two, three or in exceptional cases four instances, depending on the size of the litigation and the procedures governing the litigation. Experience shows that one instance in European courts is faster than arbitration but adding the various instances in time together, normally state courts are longer than arbitration.

It is needless to say that arbitration is recommended, especially if the parties cannot agree on one state, where litigation would be held, then arbitration is the powerful instrument to solve problems, if they cannot be solved by negotiation or by soft resolution methods as mentioned in Sect. 6.3 above.

The authors strongly recommend to negotiate and to avoid any legal proceedings. As mentioned before: A bad compromise is better than a favourable judgement.

6.4.6 The Preparation of Evidence

When discussing the tasks of the "Monitor of Litigation" we have already stressed the importance of the documentation of evidence for the litigation, if litigation should be started despite all the attempts made. Whilst it is outside the scope of this book to recommend on how the file of evidence should be prepared, we do want to draw the attention of the reader to the independent and outside evidence which in many cases is necessary.

Disputes do not always arise at the end of a project, in fact they normally arise during the execution, when the execution has to go on and the project cannot be stopped without risking heavy penalties. Therefore it is important to record evidence during the project and agree with the other party on how the evidence should be recorded. Often such agreement cannot be found, since the recording itself might already anticipate an opinion on the problems. In such case an outside expert should be sought.

For an outside expertise on facts that have occurred on site, again it would be desirable that the other party agrees to the invitation of the expert and to a division of the cost. Such agreement is rare, though. If no agreement can be reached, then the expert should be invited by one party. But by all means the other party should be invited to be present, when the expert records the facts on site, so that the other party cannot deny the evidence taken by the expert.

Even if the other party has been present, when the expert has recorded the facts, the evidence of the expert will be taken by the arbitrators or by the state judges as a "party-expertise", which reduces the power of such expertise. Therefore it is better to have an agreement on the expert between the parties. The expert can even help in a negotiation to solve the problems. By all means the expert should be asked to deliver his findings also to the other party. Sometimes people think, it is better to keep the evidence, which the expert has found, secret till they present the evidence in front of the judges or arbitrators. This is wrong!! One should always play with "open cards" since it helps to soften the stressed situation on site and can help to foster negotiations.

If the other party cannot be convinced to use an outside expert then the first party should go ahead and entrust the expert, pay the costs and keep the report for the litigation. The soft resolution methods discussed in Sect. 6.3 are of course a further step to collect the evidence.

In some countries (e.g. Germany) an official nomination of an expert can be used. It is called "selbständiges Beweisverfahren" (independent proof procedure), which can be requested to the state courts by either party in dispute. The judge will then nominate an expert who is also paid by the tribunal, although the demanding party has to make an advance on the costs for the expert. The result of the findings in such an independent proof procedure is taken in a later litigation as non-partisan proof.

We strongly recommend, if litigation is envisaged, to use an outside expert to record the evidence. Such outside expertise has always more power than a party's own employee's testimony, even if the expert is called in by only one party.

6.5 Case Studies and Questions

6.5.1 The Soft Drink Factory

A) Description of the Conflict

- A soft drink company in Spain needed a new floor in its bottle cleaning and bottle filling department. Since the cleaning of the bottles is done with water which is charged with certain caustic solutions and other chemicals to obtain a good cleaning result, and some of the cleaning water drips on the floor, the floor had to be resistant to the cleaning water and some other chemicals handled in the hall.
- For installing the new floor the cleaning and bottling machines had to be removed from the hall. Since the machines could not be moved through the doors of the hall, one wall of the hall had to be torn down and built up after the floor was refurbished and the machines were moved back into the hall.
- A contractor was called in to apply a mastic asphalt floor of 6 cm with a special epoxy lining on top. The combination of mastic asphalt and epoxy casting resin as lining was supposed to carry the weight of the machines and resist the aggressive cleaning water and chemicals, the contract price for the new floor was approx. 40 000,– €.
- After finishing the floor refurbishment the machines were brought back into the hall, the wall of the hall was closed and production started. After only some weeks of operation the lining started to peel off from the mastic asphalt, and the machines sank into the floor by approximately 1 cm, depending on the weight of the various machines. The owner of the soft drink factory claimed against the contractor and requested refurbishment of the floor and acceptance of the cost for moving and reinstalling the machines on the refurbished floor.

B) The Parties' Approach to Resolution of the Conflict

- The contractor approached the manufacturer of the epoxy lining to make good for the damage and sued him in front of a state court. The contractor lost the case since he had applied the lining with a much thinner thickness than prescribed by the manufacturer.
- The owner of the soft drink plant was pressed by the authorities to repair the floor since there was a danger that bacteria and micro-organisms would flourish under the peeled off lining, thus putting into danger the purity of the soft drinks.
- The owner, not wanting to repair the floor and thus destroying the evidence, called in an expert via an other state court, who was supposed to examine the case and give a report to the state court. The contractor firstly opposed the necessity of making an expertise, secondly, when the court did not accept his arguments, he opposed the chosen expert for unfounded grounds of partiality and thirdly brought forward numerous reasons (unavailability of the personnel, who had executed the refurbishment of the floor and others) which delayed the legal expertise. The expert was finally able to make the expertise three years after the owner had started legal proceedings.

- The expert's result of his analyses were that the floor was not suitable for the application in the soft drink factory and estimated the cost of refurbishment to apr. 250 000 €.
- When the owner wanted to settle the dispute by negotiating a compensation, the contractor had put his company into bankruptcy. The owner finally refurbished the floor at his own expense, not getting anything from the contractor.

C) Comments on Lessons to Be Learned from the Case

- The owner selected an unqualified contractor, he should have paid more for the refurbishment and get a qualified contractor.
- The owner selected inappropriate material for the floor, consulting with an outside expert would have told him that the method chosen by the contractor was inappropriate. The owner should have tested the floor before bringing in the machines back in and rebuilding the wall of the hall.
- The owner should have called in an expert to analyse the floor, immediately when the problems occurred. This expert, although considered as party witness, could have testified in court litigation. In the absence of other testimony the court would have most likely accepted the testimony.
- The owner should have tried to negotiate with the contractor, whilst the contractor was still able to pay for the damage, before closing his company for insolvency.
- The owner should have repaired the floor and saved the money for all the legal proceedings and put the money into the repair.
- Although the owner was in all his rights, which the state court finally after 5 years acknowledged, the owner did not get a single € for all his expenses and finally had to get the floor repaired by a qualified contractor. The owner had thrown good fresh money after old bad money and did not get anything for it.

Although this case is not a major international project, it has all the characteristics which occur in international projects very frequently.

6.5.2 The Alcohol Plant

A) Description of the Conflict

- An integrated plant producing alcohol, sugar and feed stuff from corn was contracted in the 1980's by a client in an East-European country. The contract was a supply contract together with supervision and performance guarantee, it was given to a consortium of West-European companies. The erection and the supply of consumables were client's obligation.
- During the execution of the contract problems have occurred and the parties made three negotiated amendments to the contract, which took account of these problems. Pre-commissioning of the plant was supposed to start on 01.01.1986. Since it was still not started in August 1986, the parties negotiated for a new starting date, which was supposed to be the 15.10.1986. The Consortium made the Client responsible for the delay because of late erection and non-contractual consumables. As pre-commissioning was not started on the 15.10.1986, an extra

delay of one month, i.e. the 15.11.1986 was granted by the Client. This date again not being respected, the Client prohibited the further work of the Consortium on site.

B) The Parties' Approach to resolution of the conflict

- The Client gave order to an international third party inspection company to inspect the works and make a report on the deficiencies of the plant. The Consortium was invited to be present at the inspection, the invitation was accepted by the Consortium. On the basis of the report of the third party inspector the Client called in an other company to make good the deficiencies, so that pre-commissioning could start on 31.03.1986. The precommissioning was witnessed by the same third party inspection company, which had inspected the plant before, the pre-commissioning was also witnessed by the Consortium.
- The client called for arbitration, in which he requested to be paid the equivalent of 11.5 Mio. € plus interests. The claim was based on in total 17 technical deficiencies of the plant, which included such items as 3% lack of alcohol production, 3% lack of total output of the plant, malfunctions of various systems of the plant and non-contractual quality of the feedstuff. The Consortium made a counterclaim requesting the equivalent of 10 Mio. € on the grounds of delayed finishing of the erection and non-contractual quality of the consumables.
- The arbitration court called in an expert to report on the various claims of the claimant and the defendant. The expert found that nine of the 17 technical deficiencies claimed by the claimant were not justified, especially the two claims on the lack of alcohol production and the lack of total production were unjustified, thus only eight of the technical claims were accepted by the arbitration court. Concerning the counterclaims of the defendant the expert confirmed that the client was late by one year in the erection of the main production facilities, that he was late by 10 months in the erection of the secondary facilities and that he was late by 7 months in providing consumables, which had the quality specified in the contract.
- The Arbitration Court conceded a minor sum of an equivalent of less than 1 Mio. € to the claimant. The arbitration lasted for apr. 8 years and cost each party more than the equivalent of 1 Mio. €.

C) Comments on Lessons to Be Learned from the Case

- The Client seems to have done everything right, he negotiated with the Consortium, when problems arose, he gave an extra month to finish the project, he gave the Consortium the chance to commonly select a third party inspector, he called in the third party inspector to record the status of the project, he made good the deficiencies he considered as such and he made a pre-commissioning run with the same third party inspector. Nevertheless, one cannot consider his legal proceedings as successful. The Client was not even paid the total of the changes he had introduced in the plant.
- What did the Client do wrong? He did not analyse the deficiencies correctly, and did not evaluate his own shortcomings in the project. The Client was late with

his own obligations, which caused the delay of the project and only some of the deficiencies were considered as such by an independent expert. The Client only saw the faults of the Consortium but did not recognise his own faults.

- It appears that it were the lawyers who had the "say" in the Client's team and not the Top Management and the technicians. In fact, the project was located in a country, which was still under communist rule, where in many cases the party had the say and not the specialists. Orders were most certainly given by the party to go for arbitration, without having thoroughly investigated the project by a Monitor of Litigation.

6.6 Conclusion to Chapter 6

Chapter 6 discusses the various existing procedures to solve conflicts in a major project. After an introduction and a summary of key words, Sect. 6.1 presents the position of Monitor of Litigation and the difference between the Project Manager and the Monitor of Litigation. The moment when the Monitor of Litigation should be installed is detected via the Early Warning System, which allows Top Management to monitor the situation in the project execution.

Once the decision has been taken by Top Management to prepare for mediation, arbitration or litigation and thus to solve the conflicts of the project, then a Strategy of Litigation has to be established by the Monitor of Litigation. The steps to follow to build up the strategy are explained in Sect. 6.2.

Before going to court or arbitration the authors advise strongly to try to resolve the conflicts with soft resolution methods like Mediation or employment of a referee. How to go about, when employing soft resolution methods, is explained in Sect. 6.3. The employment of an outside expert is recommended as well, who can help to solve differences on technical matters between the parties. The corporate pledge model, more and more used in the US, is also explained in this section.

If soft resolution methods do not solve the conflicts then the parties will normally use a state court or arbitration, if arbitration is stipulated in the contract between the parties. Section 6.4 explains what the parties should take into account if they go to court, which precautions they should take and which arrangements should be foreseen in the contract for the ultimate tool to solve conflicts, which are legal proceedings. The strong advice is given, that the parties should always try to solve their differences outside of court, even if the legal proceedings have been started.

In Sect. 6.5 finally case studies and questions on the matters discussed in Chap. 6 are presented, which help the reader to better digest the contents Chap. 6.

6.7 Questions on Chapter 6

1. ***Monitor of Litigation***
   Explain the role of the Monitor of Litigation in a major international project, which are his tasks and what experience should he have?

2. ***Arbitration***
   Explain the difference between arbitration and mediation.
3. ***Arbitration in China***
   Which is the peculiarity about nominating an arbitrator according to Chinese law? Do you know any other country, where the same rule exists?
4. ***Early Warning System***
   What is the purpose of an early warning system?, how should it be set up? Try to list cost elements of an early warning system and give an estimate of the cost for such a system for a project, you have been involved in.
5. ***Alternative Dispute Resolution***
   List the alternative dispute resolution systems, which you know and list the advantages and disadvantages of each of these systems. Are there any restrictions with respect to application of these systems? Are there any countries, which you know of, where such restrictions exist?
6. ***Corporate Pledge Model***
   Explain the Corporate Pledge Model. Is the Corporate Pledge Model only used in case of conflict or is it a standard clause? Name the advantages of a Corporate Pledge Model. Does a Corporate Pledge Model replace arbitration or litigation?
7. ***Expert***
   How do you find an expert? Who pays the expert in case of arbitration?, who in case of litigation? In which circumstances would you call for an expert?
8. ***International Projects***
   Would you use arbitration or mediation also in one country, or are these means of conflict resolution only for international projects?
9. ***Cost of Legal Proceedings***
   List all elements, which have to be taken into account, when estimating the cost of legal proceedings.
10. ***Compromise***
    Explain, why people often say "A bad compromise is better than a favourable judgement." List the elements, which lead to this saying.
11. ***Terms of Reference***
    Explain, what is understood under Terms of Reference, when are they drawn up and by whom?
12. ***Strategy for Legal Proceedings***
    List the elements to be considered, when drawing up a strategy for legal proceedings in an international project.

References

[1] International Chamber of Commerce in Paris. www.ICCwbo.org.
[2] Klowait J, Hill M (2007) Corporate Pledge – Königsweg zur Implementierung von Mediation in der Wirtschaft, Schieds VZ 2007, Heft 2, Beck C H, München.
[3] CPR International Institute for Conflict Prevention and Resolution. www.cpradr.org.
[4] American Arbitration Association, New York, USA. www.adr.org.

[5] International Commercial Arbitration Court, Moscow. http://eng.tpprf.ru/ru/main/icac/.
[6] Barth M, Johnston G (2007) Vereinbarung einer Schiedskommission als Wirksamkeitsvoraussetzung der Schiedsklausel – Zur Nichtanerkennung eines chinesischen ICC-Schiedsspruchs in Deutschland, Schieds VZ 2007, Heft 6, Beck C H, München.
[7] Chinese International Economic and Trade Arbitration Commission (CIETAC). www.cietac.org.cn/english/laws/laws_5.htm.
[8] (2008) Cost calculator providing an estimation of the cost that would be fixed by the Arbitration Court, ICC Paris. www.iccarbitration.org.

7 Expertise Contributing to Conflict Solutions

Abstract. The very important role of an expert and his expertise are discussed in Chap. 7.

In many court-cases it has been the expert who decided on the outcome of the litigation respectively of the arbitration. Judges choose to nominate an expert, when they are unable to solve important special (technical, medical, biological etc.) matters of the case themselves and it is left to the expert to analyse the non-legal details and to present his findings to the court. The court, of course, has to be convinced on the validity of the expertise, taking into account the arguments against or in favour of the findings of the expert.

Experts are used in Court or Arbitration cases to assist the judges – they can also be used by the parties in their attempt to settle their disagreement or conflict before a court or arbitration case is started (see Sect. 6.3). At court there are experts used for many areas, such as technical, medical, legal, biological, economic, consumer goods and other areas. In this book we shall only look at technical expertise, since this area is the one that plays a predominant role in legal proceedings of international projects.

The fields, in which the expert might intervene are e.g.:

- technical assessment of machinery and/or processes
- assessment of foundations, construction, quality of buildings
- testing the quality of products
- analysis of the conformity and performance of industrial plants
- investigation of damages and defects of plants, machinery, material and resources
- delay investigation
- analysis of the extra costs claimed

In Sect. 7.1 advice will be given on how to influence the choice of the expert, when selected by the Court. It is always the Court who names the expert although the parties may make suggestions to the Court.

Section 7.2 will treat the potential intervention of an expert during pre-arbitral or soft resolution methods. It is always advisable to try to solve litigation by using an independent expert who can mediate between the parties.

Section 7.3 will show how the expert intervenes, once he has been nominated by the parties or the Court. Advice will be given, what should be done and what should not be done by a party in order to come out successfully in an expertise.

If the outcome of an expertise is not successful for the party, what can be done in order to reject an expert and thus to have rejected the expertise by the Court? Such questions will be treated in Sect. 7.4.

The cost of an expertise will be treated in Sect. 7.5 and in Sect. 7.8 we discuss Case Studies and put forward questions to students, on the basis of which they can check, whether they have well "digested" the text of this Chap. 7.

Key words: Expert, Expertise, Procedure for nominating an expert, model clause for expertise, Chambers of Commerce, SVV.IHK.de, International Chamber of Commerce (ICC) Paris, Centre of Expertise, International Commercial Arbitration Court in Moscow, Chinese International Economic and Trade Arbitration Committee, International Centre of Expertise, subcontractor, VAT, Value Added Tax, Rejection of an expert, Cost Schedule for Experts, contractual penalties, commissioning, pre-commissioning, American Arbitration Association

7.1 The Appointment of the Expert by the Court

In the case of arbitration or litigation at a state court the expert is always appointed by the Court, the parties only have the right to make proposals for a specific expert. Given the importance of the result of an expertise the parties have to be very cautious during the choice of the expert by the tribunal. Much too often the parties do not know an expert and depend on the court to suggest one.

If the parties decide on arbitration in their contract drafting, then it is always advisable to define the nomination procedure for an expert already in the contract. As exposed in Chap. 6, it is even advisable to nominate one or three experts already in the contract, if the subject of possible disagreement can be foreseen at the time of writing the contract. This is often omitted by the parties, and it becomes difficult to find an agreement on the procedure for, or on the nomination of an expert, once the dispute has started. Of course, at the moment of drafting the contract, the parties often cannot foresee the kind of dispute, that might arise during execution, is a civil engineer needed?, is a financial expert needed?, or is a chemical or mechanical engineer needed?, therefore it is generally very difficult to name an expert already, when drafting the contract and therefore, all one can do, is to define the procedure of how an expert will be nominated, once a dispute has come up.

The various bodies which provide arbitration, such as the ICC in Paris, the American Arbitration Association, the International Commercial Arbitration Court in Moscow or the Chinese International Economic and Trade Arbitration Committee also offer assistance in finding experts for the specific problem the legal proceeding might require. Depending on such proposal by the arbitration body, though, carries the risk that the expert proposed by the body might not please the party requesting the expertise. The best way of having a suitable expert named by the court is to propose an expert which the party knows already. The authors have seen many trials at state-courts and at arbitration tribunals, where one party proposed an expert and the other party, in lack of an own proposal, made no suggestion, which lead to the nomination of the other party's expert by the court. It is obvious that the party which proposed the expert has thus gained a considerable advantage.

An often successful approach is to suggest the expert "unofficially" to the arbitrator, whom the party has named as member of the arbitration court. This arbitrator can then introduce the expert to his colleague arbitrators as someone well known to him and can try to convince his arbitrator colleagues to nominate the expert. It is always helpful if the expert is suggested by the tribunal, since the tribunal has the last word to say in case of opposition to a proposed expert.

Experience has shown that only physical (natural) persons should be entrusted with an expertise. It should at all means be avoided to name a legal entity as expert (a certain institute of a university, a special research company etc.). If a legal entity is entrusted with an expertise, then the actual expert is not known at the time of nomination, the person performing the expertise might change during the course of the expertise and the risk exists that management of the legal entity might intervene in the formulation of the expertise without having sufficient knowledge of the cause at stake. All this might cause dangerous results in the expertise.

For choosing the expert, the ICC Paris proposes as a model clause to be included in the contract as follows:

"The parties to this agreement agree to have recourse, if the need arises, to the International Centre for Expertise of the International Chamber of Commerce in accordance with ICC's Rules for Expertise."

It is advisable to insert the model clause for expertise in a contract in addition to a model clause for arbitration mentioned in Chap. 6, since the clause for arbitration does not necessarily entail the nomination of an expert by the body settling the dispute through arbitration. Furthermore the expert might be needed during execution of the project for settling disputes outside an arbitration or litigation.

If the proceedings are handled by a Court the expert will be nominated by the Court. Since the courts often don't have sufficient information on available experts at their disposal, they refer to other bodies such as the Chambers of Commerce, the Universities, well known research institutes or directly to independent engineers, physicians, chemists etc. Now that the internet is available, one can also find experts in the internet under various addresses. For a German expert, e.g., one can find a list of experts under www.svv.ihk.de. There some 8 500 experts are listed with their various fields of activity. Inserting the topic e.g. "Kraftfahrzeugsachverständiger" (expert in motor-vehicles) and the postal code of the area, where the expert is needed, the web will offer a list of experts from which the user can choose. Recently other private institutions have also developed lists of experts, but the list of the chambers of commerce is normally the most comprehensive. In the USA experts might be requested from the American Arbitration Association, for this the reader should look up in the internet under www.adr.org.

From the above it is obvious that the party has an interest in finding out about an expert's capabilities and experience. This is not an easy task and has to be done without too much time available, since the court will not give much more than two or three weeks to respond to the question, whether the party has objections to the expert. For important cases, though, the party should make the effort to make some inquiries. An often applied instrument is to have someone, who is not part of the party, make the inquiry through a call to the expert himself or asking the body listing the expert about his background. A party cannot argue, that they want to reject an expert, because they did not know him (her), once they have seen the report the expert has put forward.

An often discussed question is, whether the expert should be a generalist or a specialist for the specific type of project in litigation. One might think that the specialist is the right choice. If the legal proceedings are about big international projects, the authors are in contrary of the opinion that the generalist is the better choice for the following reasons:

- the specialist is often limited to his specific know how whereas a big project needs knowledge about many different technical and managerial areas
- the generalist normally has more experience in expertise for national or international courts, having been exposed to many litigation cases and has thus more knowledge about legal procedures
- the specialist might concentrate on the problems lying in his specific field of know how and treat other problems, which he has not yet been opposed to, in a very general way

An example of the above is a legal case, which had to solve the problem of an often clogging pneumatic dust conveyor in a conventional power plant. The expert called by the Tribunal (a professor from the University, let's call him Expert 1) came to the result, that the clogging was due to the wrong layout of the pipes of the conveying system, whereupon the Tribunal wanted to blame the provider of the conveying system. This provider then called in an expert witness; let's call him Expert 2, who found out that the problem was caused by a discontinuous feeding of the combustible which in turn led to discontinuous feeding of dust to the pneumatic feeding system. Expert 1, a well known professor and engineer for pneumatic feeding systems, just made a calculation of the feeding system, without ever going on site to look at the problem physically. It was the generalist, Expert 2, who saved the case for the provider of the feeding system, since he looked at the whole power plant, analysed all the systems of the plant and came to the result, mentioned above, of discontinued feeding of the

Fig. 7.1. Grain processing plant with a large number of components

dust to the conveying system. The Tribunal was convinced by him that the calculations of the specialist (Expert 1) were not correctly adapted to the power plant under consideration. The specialist had not looked over the "rim of his plate", which can easily occur, if the view of the total picture is neglected with respect to the view of the detail.

The above example shows that a party can well attack the findings of an expert, the expert's opinion is not sacrosanct. Possibilities of rejecting an expert are treated in more detail in Sect. 7.4 below.

It has to be taken into account that big international projects, such as the one shown in Fig. 7.1, have a large number of components and subcomponents which all use different technologies and which would, in case of a conflict, require a specialist for the specific technology under dispute.

In conflicts concerning big international projects it is very rare that the conflict only concerns one single component and one single technology. Normally there are a number of issues under dispute which would need a specialist each one. It is obvious that it would hardly help the court to nominate various experts for each specific issue under dispute. What is normally done, is that the court nominates a technical generalist, well experienced with the kind of plant under consideration, who in turn might call on one or the other specialist for issues that go beyond his experience. The generalist will then integrate the findings of the specialists into his report.

7.2 The Appointment of the Expert by One or Both Parties

As has been mentioned in Sect. 7.1 above, the authors suggest that the parties try to find an agreement on an expert before the tribunal appoints one. This expert can then help the parties to find an amicable solution and bring the litigation to a fast end. This is, in the opinion of the authors, the best way to solve a litigious problem and to avoid the cost attached to a formal litigation.

If the expert has not been able to find a solution to the conflict, acceptable to both parties, and the parties still feel that they have to go for formal proceedings, then the same expert chosen by both parties might also be appointed by the Court. This is, though, the exception to the rule, since one of the parties will most probably reject the expert, because his proposal was not favourable to them at first hand and gave the reason not to accept the amicable solution. Then the Tribunal may use the expert, who had been chosen before by the two parties as an expert-witness. In case that the expert is then able to convince the Tribunal, then a decision might be taken by the Tribunal and no further expert needs to be nominated and the cost spent for the expert by both parties on a 50:50 basis was not spent needlessly.

In many cases the parties cannot agree on a person to help solve their differences. Before going to Court, it is advisable that the party who intends to ask for legal proceedings finds an expert and asks him to report on his views of the case. Such report can afterwards be used as a proof and the expert may appear at court as expert witness as has been explained above.

7.3 Execution of the Expertise

The Court has to take a formal decision for the entrustment of an expert. This decision will include:

- name and address of the expert,
- the Terms of Reference for the expertise,
- and the party who requested the proof specified in the Terms of Reference.

The parties should at all means thoroughly study the Terms of Reference and if necessary request corrections.

Normally the expert does not participate in drafting the Terms of Reference, since only after they have been drafted, the type of expert needed, can be specified and be appointed. This means that the arbitrators or the judges have to formulate Terms of Reference in a discipline (technical, chemical, medical, etc.) where they have little knowledge. The outcome are sometimes Terms of Reference which have shortcomings in the wording or where questions are formulated which simply cannot be answered by the expert. This will especially occur, if a party wants to mislead the Tribunal and deviate from evidence which is disadvantageous for that party.

To overcome this difficulty, the Tribunal has the right to nominate the expert before finalizing the Terms of Reference and request the expert to participate in the formulation of the Terms of Reference, which then avoids the above mentioned difficulties. Often the expert can limit his scope of work by intervening at an early stage in the trial and by helping to formulate the Terms of Reference.

If the expert is nominated at an early stage of the trial, he can also help the Tribunal in the formulation of a potential amicable settlement. With his technical knowledge he can propose solutions which the Tribunal might then include in the draft of a formal proposal for an amicable settlement. The authors have themselves proposed such solutions in many arbitrations and state tribunals, avoiding thus time consuming trials and the costs related to them.

At state courts the expert will normally be sworn into his task, binding him to an independent execution of the expertise in line with the existing technical and scientific rules. In an arbitration such swearing-in is not the rule.

In many cases the expert will base his findings not only on the written pleadings of the parties but will physically inspect the subject at stake. Such site inspection has, of course, to be made in the presence of both parties. It is advisable that the expert invites the parties to a discussion on his findings during or right after the site inspection, where he will formulate orally his findings and have the parties comment on his findings. As a result of this discussion the expert will request further documents from the parties, which can prove their view of the matter. It goes without saying, that these documents, sent to the expert, have to be copied to the other party but not necessarily to the Tribunal.

It occurs that a party can only prove their view of the matters by revealing secrets of the company. Such secret can be a specific formula for a product, a special production method, the use of a rare material etc. The expert then has the right to request the party to reveal the secret to the expert by excluding the other party during the taking of the evidence. The secret will be treated in the expert's report secretly in the way that the opposing party gets a report with the respective portion of his report in blank and the full report will only be revealed to the Tribunal and to the party holding the secret. In all other cases the expert has to act in the same way as a tribunal, giving the two parties full access to all information he obtains and thus giving both parties equal process of law.

In the preparation of a site inspection by the expert the question regularly is raised, who may assist the site inspection. Generally speaking each party has the right to be present, and the expert has the obligation to formally invite both parties to the inspection. As for court appearances the expert is obliged to give the parties sufficient time ahead of the inspection. A lapse of minimum three weeks is considered sufficient in most legal procedural laws.

Sometimes parties try to exclude certain persons or companies from participating in an expert's inspection. The reason for such exclusion can be the secrecy of certain aspects of the project, competition, certain occurrences in the past with such persons etc., which occurs especially if a third party had to intervene in a project in order to make good the deficiencies of the other party. (See in this respect Sect. 6.5.2.)

The tribunal or the expert cannot exclude persons from such inspections if they are members of the party including the lawyers of the party (provisions as discussed above with respect to secrecy have to be made). A different matter is the participation of representatives of subcontractors of one of the parties. The subcontractor, not forming part of the legal entity of the party, might be a competitor of the opposing party, which would want him excluded from the expert's site inspection. Depending on the regulations of the procedural law governing the legal proceedings, the tribunal has the right to exclude such subcontractor(s) from the expert's mission, if the other party should so request. Needless to say that the Tribunal has to well support his decision by valid arguments.

The expert will abstain in his report and his statements from any legal analysis; he is only allowed to make assessments in his field of competence. If an expert enters into a legal analysis stating, which party is right or wrong, then he risks to be rejected by the Court or by a party as biased which will lead the expert to be removed from his task. Nevertheless the expert will sometimes come across legal questions in relation to his task, but these questions he has to relate to the Tribunal and ask for guidance how to approach the case.

An expert may never assign additional tasks to himself; he has to stick strictly to the Terms of Reference and may not go beyond these terms, even if he might find evidence that might play a role in the litigation. If, on the other hand, both parties agree that the expert also looks into a certain problem, which had not been included in the Terms of Reference, he might do so, but he should always inform the Tribunal about such extension of his task and ask for approval to elaborate on such a problem in his report.

In all cases the expert will keep close contact with the Tribunal and ask for guidance, if he has doubts about how to proceed in his work.

The expert will invite the parties to participate in the site inspection he intends to perform through the lawyers of the parties. This does not mean that the lawyers have necessarily to assist the site inspection. It is good practice that the expert leaves it to the parties whether they want their lawyers to participate to the site inspection and the technical discussions or not. Having seen the costs of an internationally experienced lawyer (see Sect. 6.4.4 above) it is well worth a consideration, whether the lawyer needs to participate in the technical discussions.

During his mission the expert may come across questions, for which he needs assistance by an expert of another discipline or by the services of a laboratory. In the case study in Sect. 6.5.2 the expert might have used the services of a laboratory to analyse the quality of the production water used in the alcohol plant. The expert has the right to utilise such assistance by subcontracting, but he should always inform the Tribunal, if he intends to use the services of an other expert or a laboratory in the course of his investigations.

The expert may use third persons to assist him in his expertise. Such third persons may be his employees or outside persons if necessary, but it is always the expert who carries full responsibility for the whole expertise, such responsibility may not be conveyed to other persons. The expert may therefore never reject responsibility for his expertise by trying to lay the responsibility on these third persons.

Once he has finished his investigations the expert will write his report to the Tribunal. This report will first state the Terms of Reference, will develop on the findings, eventually accompanied by photos, or other evidence, and will at the end give an answer to each question put forward by the Tribunal. In some rare cases the Tribunal requests the expert to appear in front of court and present his findings in oral form without writing a report.

7.4 Can an Expert Be Rejected?

Practically the only reason, why an expert could be rejected by a party, is the proof that he is biased or that he is not adhering to the procedural laws governing the proceedings. Such proof will of course be looked for by the respective party if the expert has expressed findings which are not in the interest of that party. Such findings are naturally not a reason for the expert to be rejected by the Tribunal, the party has to prove that the expert is biased or has not respected the procedural regulations, which is not an easy task. Only the Tribunal can dispense the expert from his task and it must do so, if there is sufficient evidence to have doubts on the expert's impartiality.

Proving that the expert is not sufficiently impartial needs facts such as:

- family relations between the expert and one of the parties or members of the party,
- proven friendship of the expert to members of one of the parties,
- existing business relations of the expert to one of the parties,
- involvement of the expert in the litigation through employment by one of the parties (private expertise before litigation),

- private expertise for one of the parties in former times in other matters, unless the expert has named such employment before being nominated,
- contact to one of the parties without participation of the other party (joint meals, use of automobile of a party, other advantages),
- execution of a site inspection without inviting the other party (if the other party had been officially invited to the site inspection and did not appear, the expert has nevertheless the right to perform the inspection without the other party),
- polemic, exaggerated or aggressive wording in the report,
- expert has not respected the procedural regulations in the course of his mission.

The point regarding the quality of the report can sometimes be used, if the expert was not careful in formulating his findings. Some tribunals have followed the request of a party to dispense the expert for such reasons from his mission.

The most promising way of trying to remove an expert is to prove his incompetence. There one does not need to prove that the expert is biased, but one proves only his incompetence in the subject of the expertise (e.g. lack of sufficient theoretical background or of practical experience). When a party can show by the use of an other expert, using him as an expert-witness, that the findings of the expert are in contradiction to the "state of the art" in technology or science, or that the expert has simply made mistakes in his calculations, then the Tribunal might more easily accept to disregard the expert's report. The example given in Sect. 7.1 above concerning the pneumatic feeding system in a power plant is such an example.

Even if the expert is being released by the Tribunal from his assignment, he normally has the right to be remunerated for his intervention. The only case where he will not be paid, is when the Tribunal considers that the expert has intentionally made the mistakes or when he has acted with gross negligence. Gross negligence is of course a clear reason for an expert to be rejected by the Tribunal.

Finally there is a vague possibility to get an expert released by the Tribunal, if it can be proven that the expert has not done the work for the expertise himself and has delegated the work to one of his employees without knowing himself the details and without being able to respond to the Tribunal's and to the parties' questions. Very rare cases though are known to the authors where an expert has been released for not having done the work himself. Normally the expert will have looked well into the details of the expertise to be able to answer the questions, put forward to him.

A party should never fear to try to get an expert released. He can not and will not "revenge" for such intent, since it would jeopardize his position as a neutral and professional expert by doing so.

7.5 Cost of an Expertise

The cost of an expertise is difficult to estimate beforehand. Normally the experts charge their intervention on an hourly basis. Thus the total amount of an expertise depends on the extent of work the expert has to perform. If the expert is employed by a party, the party might try to employ him on the basis of a lump sum, it is rare though, that the expert accepts a lump sum.

The amount of work for an expert is difficult to calculate, even though the expert has the scope of work to do in form of the Terms of Reference. The expert's work depends heavily on the way the Court and the parties work, they will decide how often the expert has to appear in front of the Court, which calculations they request the expert to perform and which calculations the parties have available themselves for the matters that are under litigation. Furthermore his work depends on how he can perform his site inspections. Many factors influence his work, the plant to be inspected might not be functioning correctly, changes to the plant might be necessary to restore the original configuration, which is under litigation, etc.

Due to these difficulties it is impossible to calculate the cost for an expertise beforehand; it is widely accepted practice, that the expert's work is paid on an hourly basis. The expenses he has for travelling, report writing, tests, telephone etc. will be charged on a reimbursement basis.

In international arbitration, e.g. ICC Paris and others, no fixed hourly fees exist for the experts. In accordance with his reputation and experience the expert will charge hourly rates between 100 € and 200 €. The hourly rates are agreed upon between the arbitrators and himself before the expert starts his work. The organisation managing the arbitration (ICC, or other) is responsible for the payment on the basis of an invoice approved by the arbitrators. But the arbitrators are not the debtors for the expert's payment.

In the case of an expertise for a State Court, the remuneration of the expert is normally based on a schedule which is part of a governmental law or decree. In Germany e.g. the payment is regulated in the "Justizvergütungs- und Entschädigungsgesetz, JVEG " (Law for the remuneration and compensation at Court). There the type of expertise is basis for the hourly rate, ten groups are listed with an hourly rate from 50 to 95 €/h. To give an example: an expert for welding techniques may charge 60 €/h whilst an expert for machines and production plants may charge 75 €/h.

The State Courts tend to limit the total costs for an expertise by giving a maximum total cost for the expertise when the expert is appointed. The expert may then analyse the task and comment on the maximum, when the expert feels that the maximum is too low he has to inform the Court about his estimated expected cost. Normally the Court will agree on this new estimate.

Many experts in the various countries are subject to VAT (Value Added Tax), which can be as high as 25% in some countries. The entities, which receive the expert's invoice, on the other side, are normally not subject to VAT, therefore the VAT invoiced by the expert cannot be recuperated and become cost, which has to be paid by the parties. This is true for the case, where an expert intervenes in legal proceedings in his country of residence and the entity (Court) receiving his invoice not being subjected to VAT.

If the entity receiving the expert's invoice resides in another country then it has to be checked whether the expert has to apply VAT to his invoice or not. In the European Union the expert does not add VAT to his invoice, if the receiving entity resides in another European country. The question of VAT has to be carefully checked by the Tribunal, by the parties and by the expert in order to avoid unpleasant surprises at the time of invoicing. Since the cost of an expertise in international cases can well be above 100 000,– €, the amount of VAT to be paid can easily be around 20 000,– €.

As can be seen from the above given elements, it is extremely difficult for a party to estimate the costs for an expertise beforehand. The only guidance the authors can give is that the expertise costs lie by experience anywhere between 0,1% and 10% of the total sum in litigation. Small cases at court of some 5 000 € can easily cause expertise costs of 1 000 € or even more, which even goes beyond the 10% mentioned above.

The authors are unable to give any further precise figure on the possible costs for an expertise except that in complicated matters an expertise might result in costs of 100 000 € or above. In simple cases, where only one site visit and report writing is necessary for an international project, the cost could be around 20 000,– €.

7.6 Case Studies

7.6.1 The Tire Mounting Factory

A) Description of the Conflict

- A logistic company in France provided the transport services of tires from the tire manufacturers throughout Europe to one of the main automobile manufacturers in France. In the course of the "just in time" campaign the automobile manufacturer decided to streamline the wheel mounting to his automobiles by subcontracting the following tasks to one subcontractor: the transport of the tires from the tire manufacturers to his plant, the mounting of the tires on the wheel rims and the balancing of the wheels as well as the classification of the wheels in accordance with the sequence of cars on the assembly line. The logistic company, not wanting to loose the transport business, participated in the tender for the above service job and won the job.
- The logistic company, not being a specialist in the business of mounting tires on to the rims, engaged a consultant who designed a fully automatic tire mounting plant, including the balancing and the sequencing of the wheels in accordance with the sequences of cars relayed from the car manufacturer, in form of information which tires had to be fit to which car. The plant was built on the basis of the consultant's design by two contractors, one provided the conveying system and the other the tire mounting and balancing facilities. If a wrong delivery of wheels were the cause of a stop of the assembly line at the car manufacturer's, then the logistic company had to pay heavy penalties to the car manufacturer.
- After start up of the tire mounting plant a great number of deficiencies of the plant caused frequent stops of the plant and lead to heavy penalty payments of the logistic company to the car manufacturer. The plant did not provide the availability

of 98% which was contractually agreed with each of the contracted companies, but produced only with a combined availability of 79,8%.

B) *The Parties' Approach to Resolution of the Conflict*

- The logistic company requested the two contractors who had built the plant to make good the deficiencies and raise the availability to the contractual 98%. The two companies tried to make good, made some improvements and achieved only 91% availability and the logistic company kept paying penalties to the car manufacturer.
- In view of these problems the two companies now started to blame each other for the lack of availability claiming each that their own part was near 98% and that only the combined availability was too low, and that the logistic company had made separate contracts with each company and the 98% combined availability were not contractual.
- To find a solution the logistic company proposed an expert, which she would pay but whose result would have to be accepted by the other two companies. The logistic company was willing to pay the expert, since, with a positive result, she would be able to save the penalty payments to the car manufacturer. The two companies accepted the deal.
- The expert made an extensive analysis and showed the individual deficiencies to each of the contracted companies. They made good these deficiencies and the plant finally reached a combined availability of 98,8%.

C) *Comments on Lessons to Be Learned from the Case*

- The logistic company, or more so the consultant they had used, made the mistake not to make a provision in each individual contract for a combined availability. This would have been difficult to negotiate, but such a solution was an absolute need.
- The logistic company needed a fast solution, legal proceedings would have put them into bankruptcy due to the constant penalty payments to the car manufacturer during the duration of the proceedings.
- The logistic company made an other fault, not having put a penalty payment into the contract with each company, if those did not achieve their 98% availability. Such a penalty would have speeded up the repair of the deficiencies.
- The call for an expert was the fastest and most economic solution. The expertise cost around 13 000 €, legal proceedings would have cost each company, the logistic company and the two contractors a multiple of this amount. The logistic company took a considerable risk in the solution with the expert, but seen the result, it was worth taking this risk.
- The choice of the expert instead of starting legal proceedings was a win-win solution, since the three companies together built an other tire mounting plant in an East-European company a year later.
- Often it is better to search for an amicable solution between the parties without running to Court or Arbitration immediately.

7.6.2 The Flour Mill

A) *Description of the Conflict*

- A flour mill in a Middle East country had to be refurbished from an old stone mill into a modern disk mill to produce flour of internationally compatible standards. The rehabilitation was financed and contracted by a European aid organisation, designed by an engineering company of the country financing the aid and the machinery delivery was entrusted to a machine supplier of the same country. The erection of the machinery and the civil works were executed by the recipient of the aid, the supervision of the erection and the commissioning was supervised by the engineering company.
- The machinery was delivered and erected on time so that a first test run of the mill was held 10 months after signature of the contract. All production parameters of the mill were met but the recipient of the mill rejected handing-over of the plant because of too many items on the deficiency list.
- The deficiencies were made good and a second test run was executed four months later. Again all the production parameters were met and the plant was approved by the Engineer without any reservation. The Recipient, however, refused to sign the handing-over document due to a minor condensation problem in one of the storage bins.

B) *The Parties' Approach to Resolution of the Conflict*

- The negotiations between recipient and manufacturer of the machines being blocked, the latter called for an independent third party inspector from a third country, i.e. not the recipient's nor the deliverer's country. The third party inspector was paid by the machine deliverer, who defined the task of the inspector as follows:
  - the inspector to be present during tests and negotiations between deliverer and recipient
  - to assist the deliverer's team in commissioning, testing and handing-over procedural matters
  - to act as a neutral observer and report his findings to the deliverer
  - to act as a mediating expert between the parties in case of different opinions on good international engineering, commissioning and taking over practice
  - to assist all parties in reaching a fair, amicable and fast solution in accordance with the contract and good international project practice.
- Three months after the second test run a third test run was performed under the assistance/mediation of the third party inspector. The Recipient operated the plant under the supervision of the machine deliverer. The third party inspector did not observe any relevant problems and the production rate was confirmed. The only small problem was a condensation spot as big as a hand palm, for which the inspector found the technical reason, so that the problem could be settled.
- Despite the successful test run the Recipient came up with a new "problem", where he wanted the capacity of the plant measured in a different way than it was measured in the first two test runs.

- The third party inspector intervened and convinced the Recipient to accept the plant and to continue to operate it without the machine deliverer's personnel. The question of the capacity had to be analysed separately.

C) Comments on Lessons to Be Learned from the Case

- The problem encountered in the flour mill project is typical for countries, where industry is not totally developed. The client, in this case the recipient, often wants to keep the machine deliverers as long on site as possible. This saves money on the recipient's side and gives longer time to train the recipient's personnel
- The machine deliverer took the wise decision to call in a third party from a third and neutral country to mediate between Recipient and machine deliverer. This step avoided legal proceedings and gave the machine deliverer the chance to withdraw his personnel from the construction site, thus saving considerable amounts of money.
- The cost of the third party inspector amounted to something around 13 000 €, a cost much lower than any legal proceedings would have cost the machine deliverer.
- The solution of the problem shows, that it is often helpful to find ways to solve a problem without going through legal proceedings, this of course requires both parties to "step down from their high horse" and not to insist to be 100% in their own rights. Both parties cannot be 100% in their own rights at the same time.

7.7 Conclusions to Chapter 7

The role of the expert with respect to conflicts in international projects has been discussed in this chapter, and it was shown that experts can intervene in conflicts before and during legal proceedings. Whilst the expert can be in the role of a mediator before legal proceedings have been started, he is in a role of assistant to the tribunal during the proceedings.

The expert is chosen by one or both parties if he is appointed before proceedings have been asked for. Then the parties can influence the choice of the expert. Once proceedings have started and an expert is chosen by the tribunal, then the influence on the choice and the quality of the expert is very limited. It was discussed, that in these cases the parties can only try to convince the expert, show the tribunal that the expert is not right or try to have the expert be dismissed on grounds of impartiality, if one or both parties are in disagreement with the expert. All theses approaches to dismiss the expert are difficult and have led to success only in very few cases in the past. The tribunal will normally support the expert, they have chosen.

Therefore the best approach for the parties remains their own negotiation to settle conflicts. The second best solution is the introduction of an expert to help the parties find a solution to their conflict. Only the third and mostly the most expensive, time consuming and frustrating solution is the call for legal proceedings.

7.8 Questions on Chapter 7

1. ***Appointment of an Expert***
   Who appoints the expert for a court tribunal, who appoints the expert for an arbitration? Which sources exist, where an expert could be found?
2. ***Expertise***
   Should a legal entity such as a University Institute, Scientific company etc. be nominated as an expert?. Give the reasons for your answer.
3. ***Appointment of an Expert***
   Would you rather choose a specialist or a generalist for an expertise in a technical project? What are your reasons? Please give examples from your experience.
4. ***Rejection of an Expert***
   Name reasons for which an expert could be rejected by a court. Who has to bring forward such reasons?
5. ***Cost of an Expert***
   On which basis is an expert paid and by whom?
6. ***Site Inspection by an Expert***
   Who can be present at the site inspection of an expert?, name the reasons for the presence of such persons or for the rejection to participation for other persons.
7. ***Jurisdiction of an Expert***
   Has an expert jurisdiction in a court or arbitration case?
8. ***Expert's Mission***
   Define the mission for an expert in a project, in which you have participated and where to your understanding problems could have been solved by an expert.

References

[1] Sachverständigenverzeichnis der Industrie- und Handelskammern in Deutschland. www.SVV.ihk.de.
[2] American Arbitration Association, New York. www.adr.org.
[3] International Chamber of Commerce in Paris. www.ICCwbo.org.
[4] CPR, International Institute for Conflict Prevention and Resolution. www.cpradr.org.
[5] International Commercial Arbitration Court, Moscow. http://eng.tpprf.ru/ru/main/icac/.
[6] Chinese International Economic and Trade Arbitration Commission (CIETAC). www.cietac.org.cn/english/laws/laws_5.htm.

8 Project Management Tools to Help Avoid Conflicts

Abstract. Chapter 8 is not a lecture on Project Management, but a discussion of those tools of Project Management which are especially helpful in avoiding conflicts. For other tools which are used in Project Management we suggest that the reader refers to the publications given at the end of this chapter. The main tool to avoid conflicts during execution and acceptance of a project is the detailed project plan, which should be prepared before even bidding for a project. It should be refined after the project has been obtained, in order to give the right preparation to execute the project by the Project Manager and his team. Reference is made to Chap. 2.

How to prepare a good project description and the project's objective are described leading to the detailed project specifications or as they are also called to the task statement together with the project guarantees. On the basis of the specifications the development of the very important work breakdown structure (WBS) is discussed. The work breakdown structure and the resulting PERT analysis are shown to be the key elements to manage and monitor the project.

The establishment of the budget on the basis of the work breakdown structure is discussed together with procurement procedures and cost control techniques, which should lead to the planned results of the project. Reporting to the client is shown as the essential tool for the success of the project.

A discussion of how to prepare a project risk analysis together with the main elements to be taken into account in the course of the risk analysis is given. The risk analysis being the final tool for top management to decide on "go" or "no-go" for the project.

After having prepared the detailed project plan and a "go" is given by top management the other important element then is the selection of the project manager together with his team, if he had not already been appointed for contract negotiations. This selection is the second key element to move the project to success and to avoid conflicts. How to go about choosing the staffing is discussed.

A section with case studies and questions for the reader rounds up the chapter.

Key words: Project plan, specifications, Task statement, change orders, Functional guarantees, PERT, Project Plan, WBS, Work Breakdown Structure, Responsibility Matrix, Delay Penalties, Specifications, Time Schedule, Budget, Procurement, Subcontracting, Project Reporting, Project Acceptance Procedure, Commissioning, Risk Analysis, Project Objectives, Provisions, Guarantees, Bar-chart, Gantt-chart, Critical path, Virtual delay, Task responsibility matrix, Project Manager, Organisational structure, Team building, Team relations

8.1 The Detailed Project Plan

A Project Plan is the main tool to analyze and structure a project, it is the document with which the contractor clarifies the various issues of the project, and it is the basis on which the contractor makes his offer to the client. A typical breakdown of a project plan has the elements as given in Schedule 8.1.

Projects are mostly initiated by a tender of the potential client. In some cases a project is offered as an unsolicited proposal by the potential contractor. In both cases it is highly recommended that the contractor goes through the exercise of establishing a detailed project plan. Such project plan needs a considerable amount of manpower, as will be shown in the following explanations. The costs of a project plan are therefore considerable and in consequence unfortunately often avoided by the potential contractor, who does not want to go through the thorough planning exercise. Especially if the client has issued a tender, many contractors tend to just fill in the tender documents without going through the tedious task of project planning. But because of insufficient planning the contractor risks to run into conflicts with his client while executing the project.

A project plan should normally be established at least by two persons, a technician and a controller, if the potential Project Manager is already appointed he should participate in the project planning as well.

8.1.1 Project Description and Objectives

This part of the project plan is usually furnished by the client. It describes the client's intentions with the project in detail, and should be studied carefully by the contractor. Only if these intentions can be fully accomplished by the contractor during execution, then the client will be willing to fulfill his obligations of paying the contractor. Any uncertainty in the client's objectives should be clearly analyzed and clarified with the client. Conflicts will easily arise from inaccurate and poorly defined objectives. The contractor should insist on quantified objectives whenever possible.

Special attention, to avoid conflicts, should be given to the provisions, which the client has to make for the project; many conflicts can arise from late and insufficient

Schedule 8.1. Elements of a Project Plan

1. Project Description and Objectives
2. Specifications or Task statement and Functional Guaranties
3. Work Breakdown Structure (WBS)
4. Time Schedule (PERT) and Delay Penalties
5. Organisation, Staffing and Responsibility Matrix
6. Procurement and Subcontracting
7. Budget
8. Governing Laws and Standards
9. Project Reporting (internal and external)
10. Project Acceptance Procedure
11. Risk Analysis

Schedule 8.2. Preliminary Analysis of Project Description and Objectives

1. Has the client a competent project organisation with adequate staffing?
2. Has the client sufficient funds to finance the project?
3. Are the client's supplies realistically available?
4. Are the objectives realistic and are technical solutions available for reaching them?
5. Is the overall timeframe realistic and not too ambitious?
6. Are sufficient and experienced subcontractors or project partners available for all the tasks to be outsourced?
7. Are political obstacles to be expected in the country of execution?
8. Are there any legal constraints influencing the project?
9. Are strikes, port problems, logistic shortcomings, personnel shortcomings, lack of energy, lack of raw material, problems in infrastructure etc. likely?
10. Are sufficient in house funds available to provide the working capital for the project?
11. Is sufficient know-how available in-house?
12. Has the technology of the project proven successfully, is it compatible with the standards of the country of execution?
13. Are the guarantees and bonds to be given realistic and available, are there any currency risks?
14. Will the project most likely yield a profit?

provisions by the client. The contractor should thoroughly analyze during project planning, whether the provisions by the client are realistic, sufficient, of the quality needed and foreseen to be delivered when the project needs these provisions. It is prudent to look, whether alternatives are available, in case the client is not in a position to deliver.

Further questions to be asked in view of avoiding conflicts potentially arising out of the Project Description and the Objectives are given in Schedule 8.2.

Before pursuing in making a detailed project plan as suggested in numbers 2 to 10 in Schedule 8.1, Top Management of the company has to request a first tentative answer to the questions of Schedule 8.2. There is no reason to pursue the project unless all questions of Schedule 8.2 have given positive results. The Preliminary Analysis will implicate Top Management at an early stage of the project and raise their sensitivity; most of all it will clarify many points with the potential client and thus help avoid conflicts at a later stage.

8.1.2 Specifications or Task Statement and Functional Guarantees

In the Specifications (sometimes called Task Statement (Eisner 2002)) the client specifies, what has to be done in the project and how. It goes without saying that here the contractor has to analyse the specifications line by line with respect to the questions which have been listed in Schedule 8.2. Along with the specifications the client will also stipulate the functional guarantees the finished project will have to fulfil. Special attention has to be given to the wording and the quantification of the functional guarantees, the method, how these quantified guarantees have to be measured, have to be precisely defined as well. Payment will depend on the fulfilment of the functional guarantees, therefore the wording and quantification have to be precise. Many conflicts and litigations have resulted from insufficient precision in the definitions of the guarantees (see Case Study in Chap. 7.

With respect to tasks and related guarantees deviations will always occur during execution of a project; a planned machine is no more available on the market, the soil conditions differ from the planning, the environmental laws have changed, the client has changed his mind about his provisions, the raw material foreseen as input to the plant is of different quality, etc., etc. The reasons for deviations from the project plan are many. There is no project, which has been finished exactly the way it had been planned. Conflicts will easily arise out of such deviations. The contractor sees a chance to improve his contractual conditions (mainly price) and the client will try to reduce the price because of unforeseen deviations. Such conflicts have to be solved by applying the proposals which the authors have given in Chap. 3 and a confident relation between client and contractor is always a must.

Besides such confident understanding a procedure on how to handle deviations or changes to the project is an absolute must in any contract. A procedure of how to handle the so called "change orders" has to be negotiated at an early stage between client and contractor. Having established such a procedure does not avoid conflicts with respect to project deviations, but it provides a format of how to approach the change orders.

When the functional guarantees are analysed and concluded between client and contractor, immediately the question of penalties for not respecting the guarantees has to be agreed upon as well and included in the mutual contract. But not only the guarantees and the penalties have to be defined, but also the procedure of how to prove the fulfilment of the guarantees has to be established. A clear definition of acceptance testing, pre-commissioning, and commissioning of project elements of whole sections and the whole project is a must in every contract. Many conflicts can be avoided if these tests are well defined in the contract. (See Sect. 8.1.10 below for Project Acceptance Procedures.) The reader might be surprised how many contractors and clients have deviated in how to prove a certain output, input, noise, energy consumption, man-power need etc. at the end or during a project. Testing and commissioning are always a major source of dispute in a project.

8.1.3 Work Breakdown Structure

The work breakdown structure exposes and illustrates the different works that have to be accomplished during the execution of the project, it is the document to which many other documents such as a material provision schedule, a time schedule, an expenditure schedule, a testing schedule etc. will refer to in the project planning. The work breakdown structure splits the total work into sections, subsections, elements, sub-elements, etc. Figure 8.1 shows such a structure for a desalination and power plant. Obviously the work breakdown structure of such a big international project has a size which can not be reproduced here. Figure 8.1 just shows the structure and one possibility, how the work could be structured. The dotted lines show that more sections/subsections have to be added.

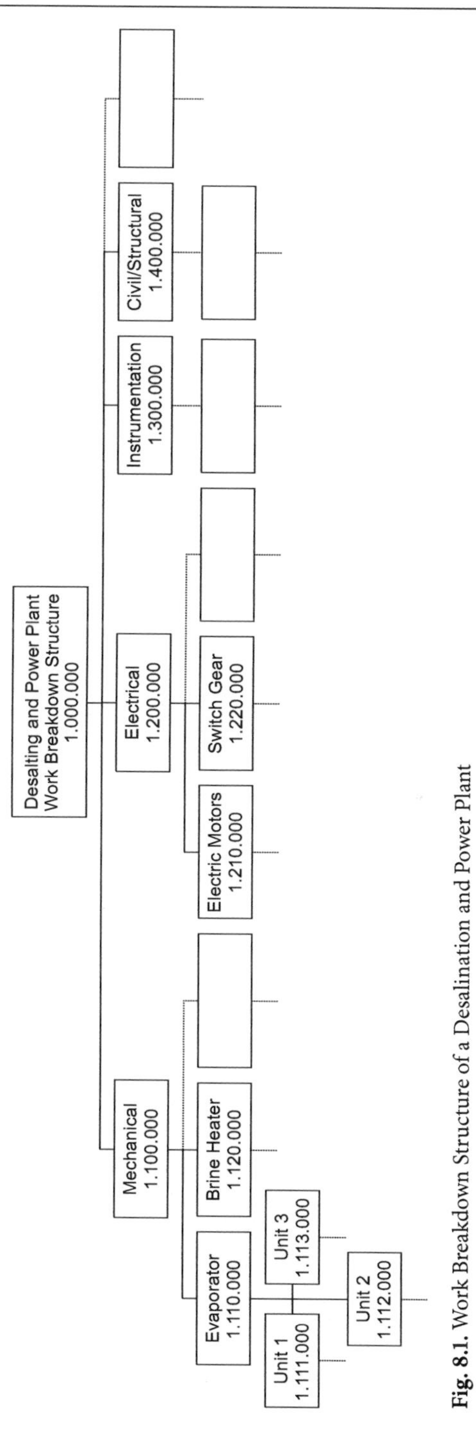

Fig. 8.1. Work Breakdown Structure of a Desalination and Power Plant

A work breakdown structure includes the number of each work-package, so reference can be made via these numbers in other structures such as a PERT-structure (see Sect. 8.1.4). Normally the work breakdown structure is not drawn up in the way shown in Fig. 8.1, the size of paper, it would have to be drawn upon, might cover a big wall. The work breakdown structure is drawn up with the adequate computer-program and printed out in a consecutive manner according to the numbering system, in a similar way as shown in Table 8.3.

The work breakdown structure should be in total congruence with the task statement (see Sect. 8.1.2) of the client. Any discrepancy with it can lead easily to conflicts during execution of the project. If the client's task statement is less detailed than the contractor needs it, then the work breakdown structure should be congruent with the task statement as far as the task statement reaches the detail. If the client has issued a work breakdown structure, which is only very rarely the case, then the contractor should stick to the client's structure. Any deviation is a potential source for conflicts.

Each work package of the work breakdown structure has to be well defined and the interface to other packages has to be described in order to avoid misunderstanding. A badly defined work package can easily lead to discussions on payment obligations by the client. One might say that such unclear definition will be made good by the other work package, which is somewhat related. This is a bad approach, though, since the discussion on the completion of a work package is unnecessary if the package is well defined and the progress payments will be paid more easily by the client.

The task statement and thus the work breakdown structure are the basis for the client's payments. The client will pay in line with the accomplishment of each work package. As will be shown in Sect. 8.1.4, the scheduling of the project is also based on the work breakdown structure. Payments for work packages will also depend on their timely accomplishment. Delays are in most international contracts subject to penalties.

8.1.4 Time Schedule (PERT) and Delay Penalties

Whereas the work breakdown structure shows the physical interdependence of work packages, the time schedule shows the interdependence of work packages with respect to execution and time, it shows, which work package(s) is (are) prerequisite for the start of an other work package. As a simple example we look at the construction of a house, where the work package "foundations" has to be finished before the work package "erection of the house" can be started. See Fig. 8.3, which shows the interdependence of work packages for the erection of a prefabricated family house.

The time schedule does not only show the logical interdependence of the work packages but it also shows their necessary time for execution as well as the starting time and the finishing time of the activity. Two main presentations of a time schedule are in use in project management: The Gantt Chart and the PERT Chart. Both presentations are used in parallel in projects, the Gantt Chart for a less detailed presentation of the schedule (mostly for top management presentations and presentations to the client) and the PERT Chart with the detailed program evaluation for project control during project execution.

The Gantt Chart or Bar-chart, as it is often called, shows the execution time of a work package (activity) represented by the length of a bar against a time line and positions the work package with respect to a time axis. Figure 8.2 shows an example of a bar-chart. This bar-chart is a graphical printout of the tabular printout given in Table 8.2. As can be seen the bar chart does not give the earliest and the latest dates, it just gives one date for the beginning and for the end of an activity. In this case the bars represent the earliest dates, as can be seen by comparison of Fig. 8.2 with Table 8.2.

The bar-chart presentation can show dependencies, which means, that it can show whether the start of an activity depends on the execution of an other activity. This is indicated in Fig. 8.2 by the arrows pointing from one activity to the other.

The PERT Chart shows the starting time, the finishing time and the work packages which have to be finished before a new work package can be started. If it is presented as a chart, like the one shown in Fig. 8.3 it also includes a time axis as abscissa. Mostly, though, the PERT Charts are presented in tables showing the registration number, the description of the work package, the length of time for execution, the earliest starting and finishing time as well as the latest starting and finishing time, furthermore it shows the slack available for the activity.

Figure 8.3 shows the PERT-chart for the Erection of a prefabricated family house. The related necessary activities to erect the house with garage and garden and the related times for execution of the work are given in Table 8.1. These activities are shown in Fig. 8.3 in their logical relation. The circles in Fig. 8.3 represent the events and the arrows between the events represent the activities, which have to be executed in order to arrive at the events. For example at event 2 the earth works have been finished, in order to arrive at the event "foundations finished" the activity "build foundations and cellar" has to be executed.

The position of the events with respect to the abscissa shows the time when these events occur. The dashed arrows in Fig. 8.3 are dummy activities, they are not executed and are only introduced to show a dependency, for example event 4 "Founda-

Table 8.1. Activities for a Prefabricated Family House

from Event	to Event	Activity	Execution Time [days]
0	1	Perform Land survey	2
1	2	Excavation earth works	4
2	3	Install Utilities (sewage, electricity, water)	5
3	4	Dummy	0
2	4	Build foundations and cellar	14
4	5	Erect prefabricated house	7
5	6	Finalization of earth works	2
5	7	Finalization of the interior of the house	5
6	8	Foundation and erection of garage	2
6	9	Plantation of the garden	5
7	11	Quality check and handover	2
9	10	Install sidewalks and pavements	5
10	11	Dummy	0

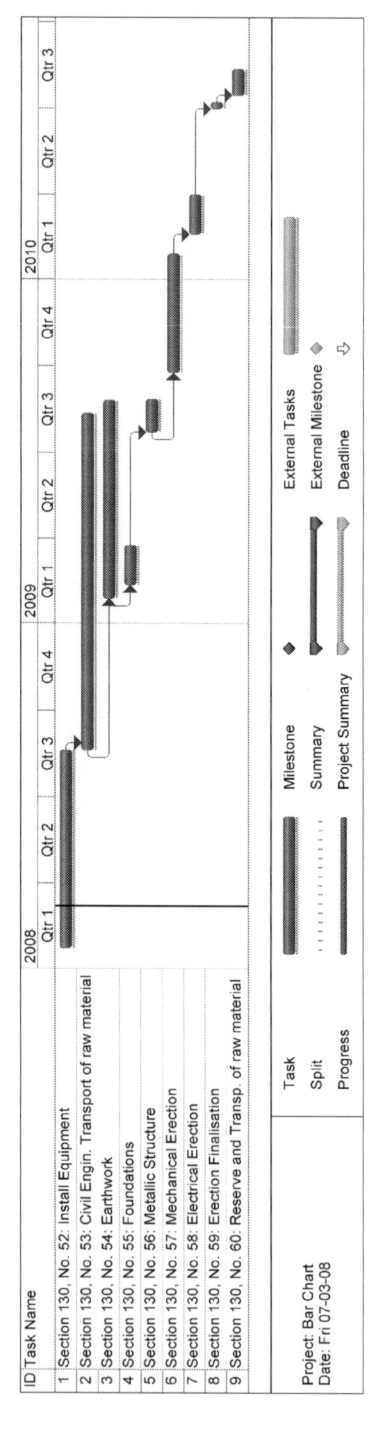

Fig. 8.2. Typical Example of a Bar Chart (Gantt Chart)

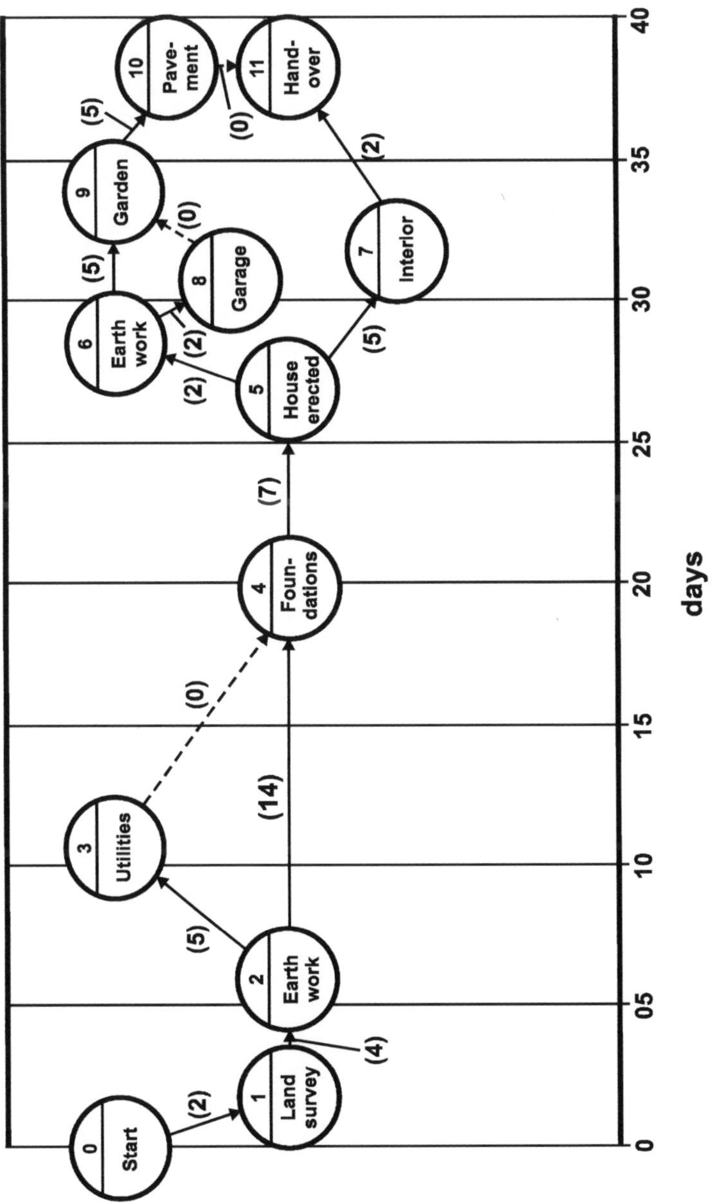

Fig. 8.3. Example of a PERT-Chart for the Erection of a prefabricated Family House

Fig. 8.4. Raw Material processing plant

tions finished" needs as a prerequisite the installation of the utilities, which is represented by the dashed arrow from event 3 to event 4.

The scheduling of big projects, such as the one shown in Fig. 8.4 of a raw material processing plant needs a computerised PERT-analysis. With such programs the earliest and the latest finishing date can be calculated and in case that the finishing date does not meet the requirements of the client, i.e. that the project end is too late, then the computer will show the so called "critical path", which is the longest path through the network. If the project has to be finished at an earlier date, then the critical path has to be analysed, whether there are activities, which can be shortened by introducing more machines, personnel etc.

It can easily be understood that a PERT-chart of the type shown in Fig. 8.3 would take very large dimensions, if it were drawn up for a big international project, such as the one shown in Fig. 8.4. For such projects the PERT-analysis is handled by special computer programs, where all the events, the activities with their related execution times and the interdependency of the events are entered as data. The result of such an analysis is shown in Table 8.2 for the raw material transport unit of a process plant. It shows the activities with their related numbering, their lengths in weeks and the earliest and latest starting and finishing dates for each activity.

The PERT evaluation technique allows to analyse the project with respect to a given end date of the project. The earliest end date is the date at the end of the critical path, which is the string of those activities which form the longest path through the network and which has no slack. It is called critical, since a slip in execution of any activity on the critical path will directly delay the end date by the slip occurred. In Table 8.2 we see that activities 130–52, 130–53 and 130–54 have no slack, any delay in one of these activities will immediately delay the finishing date of the project. Activity 130–55 has a total slack of 12 weeks, whereof 11 weeks are free slack. A free slack

Table 8.2. PERT Printout for the Construction of the Raw Material Transport Unit of a Process Plant (simplified)

Section	Number	Description	Length (weeks)	earliest Dates (weeks)		latest Dates (weeks)		Slack total free (weeks)		Comment
130	52	Install equipment	30	1	30	1	30	0	0	no slack
130	53	Civil Engin. Transp. of raw material	51	31	81	31	81	0	0	no slack
130	54	Earthworks	30	54	83	54	83	0	0	no slack
130	55	Foundations	6	85	90	97	102	12	11	
130	56	Metallic Structure	5	79	83	92	96	13	0	
130	57	Mechanical Erection	18	70	87	83	100	13	0	
130	58	Electrical Erection	6	84	89	97	102	13	12	
130	59	Erection Finalisation	1	102	102	103	103	1	0	
130	60	Reserve and Transp. of raw material	4	103	106	104	107	1	1	

can be used in extension of the activity without influencing the critical path and thus the finishing date.

Many publications and various computer programs exist on PERT techniques, we therefore abstain here from a further explanation of this technique. We shall rather look at the influence the scheduling has on potential conflicts between client and contractor.

Table 8.3 gives a listing of reasons that have in the past led to conflicts in international contract execution. The table has been established on the experience of the authors in Project Management and in dispute settling of the various kinds. It is an

Table 8.3. Reasons for Conflicts in Project Execution

	Reasons	Percent Occurrence
1.	Delays in execution	23%
2.	Discrepancies with respect to the task statement and functional guarantees	11%
3.	Problems concerning the technology used for the project	5%
4.	Change of the legal environment in the country of execution	3%
5.	Procurement problems in the country of execution and importation problems	4%
6.	Quality, quantity and date of client's supplies	8%
7.	Missing or unqualified personnel, most of all the Project Manager	4%
8.	Force Majeure, difference in interpretation	2%
9.	Payment interruption by client	8%
10.	Quality of execution of work	9%
11.	Accidents on the construction site	3%
12.	Differences in interpretation of the contract	14%
13.	Others	6%

estimative table, which gives an indication but it does not represent an exact statistical evaluation. This Table 8.3 helps the reader to concentrate on the most important issues which can avoid conflicts. Late finishing of a project is the most important cause for conflicts.

The time management (follow-up on progress and milestone compliance) during a project should be done by the controller and not by the project manager. If the project manager is also in charge of the time control, he will normally tend to keep a slack rein on schedule control, since he will always be absorbed by the day to day problems of the project. The controller, with respect to his control function, has to be independent of the Project Manager.

A delay control is not a tool, which is established once at the beginning of a project and then left alone. The PERT-type delay control has to be updated continuously, at least once every 2 weeks, in timely very tight projects it should be updated every week. The existing computer programs help to reduce the tremendous effort that is necessary for the time control, but it is worth the while. The reader should have a very close watch on the time control, when she/he is responsible for the execution of a project. This means that any deviation of an activity with respect to the existing plan has to be introduced into the PERT program. Such deviations can be starting time, execution time, cost, material and execution tools.

Since the delay is a major reason for conflicts, we want to mention the problem of "virtual delay". Many conflicts have been brought to litigation or arbitration, where delay, especially the virtual delay, was the reason for the conflict and the discussion evolved, who was responsible for the delay. The discussion could develop between client and contractor or between contractor and subcontractor, who ever was supposed to be responsible. In some cases, when the delay occurred with the contractor he would try to find a reason with the client which caused the delay, the same would do the subcontractor with the contractor. If the client for example has delayed the start of a contractor's activity by five weeks, allowing the start only in week 47 instead of week 42 and the contractor started his activity exactly five weeks later, i.e. in week 47, then of course the client is responsible for the delay, if the contractor has finished his activity within the foreseen execution time for the activity, for example in week 53.

If on the contrary the contractor would only start executing the activity seven weeks after the planned start, i.e. in week 49, then a portion of the responsibility would be with the contractor. How would the responsibility then be split between client and contractor? The contractor would argue that he would only be responsible for two weeks but the client would argue that he, the client, was not at all responsible for the delay, since the contractor was not ready any earlier than in week 49, so the delay of the client had no impact at all. The client would say that his delay was only a "virtual delay", which had no impact on the termination of the project. Judges and arbitrators have decided in a non consistent way with respect to virtual delay. Some have not taken into account the virtual effect, others have. We advise, therefore, the project manager, never to rely on his right to defer his activities if the client is in delay. He should do everything to start his activity at the moment the client has finished his foregoing activity. Otherwise any later delay might fall partially or totally under the responsibility of the contractor.

Delays are often very expensive in a project, since the opponent will not only apply the delay penalties but he will also try to obtain consequential losses, which can be more expensive than the actual delay penalties. Further to these costs the contractor has to bear his own additional costs due to the delay.

8.1.5 Organisation, Staffing and Responsibility Matrix

The organisation of a project is an internal matter of the contractor which is of high importance for the successful execution of a project. The right staffing with relevant skills and the qualified personnel will of course help, to successfully manage a project. With respect to avoiding conflicts two aspects are important in the organisation and staffing of a project: the qualified project manager and correct assignment of tasks to people in a task responsibility matrix. Detailed suggestions for the choice of project managers are given in Sect. 8.2.1. The task responsibility matrix allocates the responsibility of persons to tasks and should also define the interfaces between bordering tasks, so that there is no possibility of making others responsible for problems that oneself has to stand up for in the project.

By drawing up the task responsibility matrix, the organisers of a project are forced to define the work force necessary to fulfil a certain task. Needless to say, that by defining the work force in the task responsibility matrix the time of execution for the tasks are influenced, thus this matrix has an influence on the scheduling and the critical path mentioned in Sect. 8.1.4. Therefore the task responsibility matrix and the time schedule have to be worked out in an iterative process.

Correct staffing and a well defined organisation makes the project team strong and effective with respect to the client and thus avoids conflicts. It has unfortunately to be noted, though, that sometimes internal conflicts of the project team lead to a difficult relation with the client. It has even been noted in some occasions that collaborators of the project team carry their internal differences to the client and try to solve these differences with the help of the client. This strategy is extremely dangerous and has to be avoided by all means. The client might use the internal differences to impose his strategy on the project team of the contractor. It is not impossible that such internal quarrels can lead to important conflicts between client and contractor.

8.1.6 Procurement and Subcontracting

In Procurement and subcontracting the contractor slips into the role of being client (buyer). Therefore, all that has been said on the client/contractor relation, is normally valid for the contractor/subcontractor relation and thus does not need to be repeated here.

A word should be said on the often used back-to-back contracts, where the contractor makes the same or a very similar contract between himself and his subcontractor. Terms in the contract like "Terms and conditions of the main contract with client XYZ apply to the present contract." are dangerous and should be carefully analysed before including such terms in a contract. There are many areas which may eventually cause conflicts between contractor and client or between contractor and subcontractor. Such areas can be among others: confidentiality, financing, guarantees, applicable law etc. The authors therefore warn the reader to apply utmost care when making back-to-back contracts (see also Sect. 3.3).

8.1.7 Budget

The budget for a project is naturally one of its most important aspects, since for the contractor it is essential with respect to the success of the project. In the context of this book however we will not discuss the related issues here, except for one, which is often a source for dispute between client and contractor. We refer to an inconsistent and unrealistic budget-plan. The budget-plan is the basis for the payment-plan, according to which the contractor is paid by the client.

The budget should be established on the basis of the work breakdown structure. As the PERT-program has established time elements for each work package, the Cost-program assigns the cost elements to each work package. When PERT- and Cost-programs are combined, cost planning and cost control becomes a valuable tool for project management and the cost can be controlled and managed as the project advances.

For internal analysis maximum, expected (most likely), and minimum cost should be assigned to each work package. This allows management to analyse maximum, most probable and minimum result of the project. Today computer programs allow these analyses in a convenient matter and we abstain here to go into the details of such programs.

The important issue with respect to avoiding conflicts with the client is a clear agreement on the cost for each work package, which will allow payment by the client, once the work package is accomplished. Regular reporting and joint inspection of each work package by both, the contractor and the client, ensure a joint understanding of the payments to be made.

The cost of the work package has to be clear with respect to all the elements included in the cost. Such elements, besides the actual execution cost for the package, are: overheads, exchange rates, customs duties, delay penalties, increase in labour wages, taxes, insurance, any kind of retainers, permits and licenses, etc. The in-house cost calculation will normally add this kind of cost as a percentage on the total project, which of course is of no concern for the client. Therefore the work-package cost to be agreed upon with the client has to include all these elements. Many projects have run into disputes because of insufficient precision of the elements included in the cost of a work package.

8.1.8 Governing Laws and Standards

The governing laws (inluding tax laws) and standards of a project can heavily influence the success of a project. Since this book is not intended to give legal advice, we shall not venture into this field, but strongly advise the Project Team to get all possible information on the legal side from a lawyer experienced in the laws governing the contract. Advice from not only one lawyer is recommended, especially in international projects. (Reference is made to Sect. 2.5, 2.6 and 3.5.)

On the technical side many project-contracts refer to standards in very general terms. A typical wording is: "The contract is based on the Technical Standards, Specifications, rules and Codes of ... (the country of contract execution) or equivalent." But caution is necessary. Since many of the big contractors would want their technology to prevail, they normally rely on the term "or equivalent" and would want the standard of their country of origin to be applied. Many conflicts have arisen from the term "or equivalent" and many expertises have had to be made to solve the question, whether the standard was really equivalent. The contractor is advised to carefully study the standards of the country of execution and to accept, maybe with little technical changes, the standard of the country of execution.

As an example we give the case of a particle foam machine delivered from the United States to an East-European country. In the purchasing contract the CE-conformity was required. When the machine was delivered the client found out that all the cabling used in the machine was according to US-Standards with a UL (Underwriters Laboratory) certificate. Needless to say that the machine worked correctly and delivered the products as per specifications of the contract. The client requested all the cabling to be changed to CE-Standards or to reduce the price of the machine.

Since the vendor and the client did not get to terms, an arbitration procedure was launched. The judges, together with a technical expert, had to decide, whether all the cabling had to be changed, a reduction in price was justified or whether a simple table giving the equivalence of each type of cable between the US-Standard and the CE-Regulation could solve the problem. Client and vendor finally agreed outside the legal proceedings that the vendor would pay a certain sum for compensation and the "equivalence table" would be furnished by the vendor.

8.1.9 Project Reporting

Besides the day to day contact between the crews of client and contractor, the reporting is the main communication means in a project with which the relevant data is exchanged. Good reporting is the basis for a good client/contractor relationship. The flow of information is in principle from the contractor to the client, with the client reacting on the reports, if differences from the contract terms are observed by him.

The form, extent and period of reporting to the client is normally prescribed in the contract. Before signing the contract, the contractor should carefully discuss and define the degree of reporting with the client. The line to follow here is: "Not too much and not too scarce". A well informed client is normally a good client, who will not tend to look for conflicts.

In order to be able to have good reporting to the client, the Project Manager needs to get all information necessary from his organisation on site and from his home office. A project is a complex system which needs a constant and well structured information flow from the various groups involved, i.e. department for planning, for engineering, for human resources, for cost control, for production, for sales, for maintenance, for logistics, for informatics, etc. The project manager therefore has to build up a congruent information system in his project plan, which gives early

Schedule 8.3. Typical reporting schedule for a major plant in the Middle East

1. Design report with statements on drawing, calculations and proposals
2. Report on material procurement with dates and details of orders placed
3. Report on shipment of equipment with date of arrival on site
4. Report on site erection with segregation of civil, mechanical, electrical and instrumentation with percentages of completion in comparison to contract dates
5. Special report on delays, with actions to make good the delay
6. Report on personnel (labour and supervisors) and erection equipment on site
7. Progress report summarizing the progress in total and since last report with a summary in local language
8. A weekly special site report with the progress achieved during the outgoing week and any major occurrence on site
9. Report on budget control with invoices of the month

warnings of any deviation from the plan. The skill is here again to have a system which avoids tons of paper and thousands of computer files, which overload the system and nobody is able to read on the one side, and not to have too little information which gives the acting people the impression of not being well informed. A well defined and well graded information system is necessary. Experience shows that project managers give too little attention to this extremely essential part of the project. A good internal information system is the basis for good reporting to the client, and a need to avoid conflicts.

For international projects a clear agreement is necessary in the contract regarding the language to be used in each report. The translation of reports is expensive, can lead to misunderstandings and should therefore be avoided as much as possible, but made as much as necessary.

A typical set-up of a project report to the client is given in Schedule 8.3. In this case the period was set to one report per month, handed in on the 15th of each month. The items in Schedule 8.3 are to be extended if special sections of the project need special attention. Such sections can be technologically critical sections, sections with a special influence on the time schedule or sections, which need special political attention.

8.1.10 Project Acceptance Procedure

In order to avoid any conflict at the end of a project, the project plan and the contract should contain a well defined procedure for the acceptance of the sections and the total project. A typical schedule for tests and inspections is given in Schedule 8.4.

The Project Manager during contract negotiations has to give his utmost attention to the commissioning procedure in order to avoid conflicts at the end of his project. The authors have observed, that clients often try to defer commissioning, although the plant has been totally finished (some minor punch list points might be left). The reason for deferring are manifold, one major reason though, is the fact that the contractor has to carry the cost for running the plant as long as commissioning

Schedule 8.4. Typical Testing Schedule for a Desalination Plant

A) Precommissioning Tests and Inspections
   1. All tests and inspections during construction shall be performed by the Contractor and witnessed by the client
   2. All tests and inspections have to be performed in accordance with procedures approved by the client
   3. The Contractor has to maintain all necessary records, data sheets etc. of the tests
   4. The following tests and inspections shall be performed prior to completion:
      - Installation checks of equipment and systems
      - Hydraulic, hydrostatic and pneumatic tests
      - Instrument and control calibration
      - Operational checks of the equipment
      - System checks (verification of interlocks, alarm points, manual and automatic operation)

B) Tests on Completion
   1. Thirty Day Reliability and Performance Test
      - This test shall consist of 30 days of operation of the plant without failure or interruption of any kind
      - Within the 30 days, 10 consecutive days shall be chosen for performance tests, where the guaranteed output and raw material consumption has to be proven
      - During the performance test the average guaranteed utility consumption has to be proven
      - The Contractor has to prepare and get approved by the client the testing procedure (recordings, measuring points, diagrams, etc.)
      - At the conclusion of the 30 days the plant shall be shutdown for a maximum of 48 hours for inspection, to prove that continued use is possible
   2. Start-up Tests
      - After inspection the Contractor shall demonstrate two consecutive start-up cycles
      - During the start-up cycles the Contractor has to demonstrate, that the plant is capable of duplicating and maintaining normal operation for a minimum of twelve hours after start-up

has not been completed. A further reason is, that the client has more time to get his employees acquainted with the plant for a longer time, which also will save money and problems to the client. We advise that, at all means, time limits have to be fixed to the duration of the tests and to the time the client has to accept final commissioning.

The reader may refer to the Case Studies in Sect. 5.15.3 and 7.6.2 where the final acceptance was denied although the plant had proven its performance in all the parameters, which were stipulated in the contract.

8.1.11 Risk Analysis

Risks are manifold in a major international project. Much literature exists about how to foresee and thus to avoid risks in a project, see [1], [4] and [6] for example. It is of utmost importance that the Risk Analysis be made by a team of three, as pointed out in the beginning of this chapter: A technician, a controller and the future project man-

Schedule 8.5. Guidelines for a Project Risk Analysis [4], [1], [6]

A) Technical Risks
- New technology?
- New technology proven in tests?
- Software available and proven?
- Operation in the country of execution possible?
- Adequate environment for the technology?
- Equipment available or timely to be produced?
- Can all the warranties be met?

B) Schedule Risks
- Customer deliveries on time?, right quality?
- Has the scheduling been correctly performed?, which are the uncertainties in the scheduling?
- Are alternatives available, if technical, software, or other problems occur?
- Are there any dependencies outside the project?
- Is the customer experienced with similar projects?
- Is there a proven client-contractor relation, or is this relation new?

C) Organisational Risks
- Is the necessary personnel available?
- Has the client competent people for the control of the project?
- Is the project team adequately organised and competent?
- Is the project organisation compatible with the client's organisation?
- Is the infrastructure at the execution site appropriate?
- Are there legal risks?
- In case of a joint venture, is the organisation of the partners adequate?, are the interfaces between the partners sufficiently defined?, are the responsibilities of the partners correctly defined?
- Are there any political, religious, racial or similar constraints?

D) Financial Risks
- Is the client financially solvent for the project?
- Can all the necessary bonds and financial guarantees be obtained?
- Are there taxation problems?, can sufficient insurance be obtained?
- Are the costs of the project sufficiently analysed (wages, procurement items, wages in the country of execution, etc.)?
- Are there customs problems to be expected on importation?
- In case of a joint venture are the partners solvent?, can they contribute to the bonds and guarantees to be given?

ager. It is necessary that the project manager knows the risks, which he will encounter during the execution. The technician and the controller might not be involved in the execution of the project, but they have to be extremely knowledgeable in their field of responsibility.

Schedule 8.5 gives an overview, of which kind of risks have to be expected in a major project. The schedule can be used as a guideline for analysing the risks. In the scope of this book we shall abstain from going into further details of risk analysis, given that plenty of literature is available on the subject.

Fig. 8.5. New Technology: Toll System for Trucks on German Motorways

It is obvious that the questions given in Schedule 8.5 are interrelated, meaning that a technical risk can cause a scheduling risk or a financial risk or a warranty risk, etc. Therefore the analysis has to be made in an iterative manner, reconsidering the various questions once they have all been looked at.

We would like to draw the reader's attention to the fact that these risks are not only related to international projects. Risks are also common in national projects, maybe with some aspects not being present in national projects, such as a foreign legal system, language problems, customs, unknown national habits etc.

As a typical historic case we can cite the introduction of the Toll System for trucks on German motorways. Figure 8.5 shows one of the measuring units on a motorway, which has to check, whether a truck has paid his toll before entering the motorway. No toll stations, where the trucks have to stop and pay, are installed anywhere, so the system totally relies on the information obtained from the measuring bridges (see Fig. 8.5), whilst the truck moves on the motorway underneath the bridge at his normal speed of 80 km/h.

The technology was new at the time. The project was a disaster for the consortium, which obtained the project from the German Government. The main occurrences where:

- the technology did not work at the beginning
- the software of the whole project was delayed and with major faults at start up, the software had not been tested sufficiently
- the system was not well analysed, e.g. trucks taking toll-free roads, by-passing the motorways
- the management structure (who does what) was not clarified between the joint venture partners
- the client (the German Government) did not have clear organisational structures on their side
- no alternatives were available in the moment of crisis

At the end the client (German Government) made a claim against the joint venture for around 4 billion Euros. The arbitration is ongoing at the time this book is being published. The revenues per year of the German Government from this toll system are around 3.5 billion Euros per year.

8.2 The Personnel for the Project

From the discussions of the detailed Project Plan in Sect. 8.3 we can conclude that, as is obvious, the personnel is the decisive factor with respect to the project's success. The Contractor therefore has to take utmost care in selecting the personnel for the project. In the following sections we make proposals, how to proceed in the selection of the Project Manager, the decisive person in the project team, and how his team should be built up, selected and organised in order to form an efficient team.

8.2.1 The Recruitment of Project Managers

The importance of the Project Manager has evolved considerably in the last years. In the 1960's a project manager was considered more as a controller with a lot of administrative work for a project. He was not really considered a manager in the sense of creating the plant, the bridge, the highway which was the object of the project. This attitude towards a project manager has changed, nowadays the project manager has a higher esteem in the company of the contractor and normally if the project is of an important size he reports to top management. Figure 8.6 shows an organisational chart with project management reporting directly to top management.

The direct reporting of Project Management to Top Management ensures that the Project Manager or the Project Director, if more than one projects are run at a time, is of the same level as his colleagues in Engineering, Production etc. This helps Project Management to be supported by the functional managers Engineering, Production etc.

The organisational structure outlined in Fig. 8.6 is also helpful in relation to the client. A Project Director reporting directly to Top Management is respected by the client, and can negotiate at eye-level with top management of the client. The Project Director or Manager does not need to get approval by various functional divisions but he can get approval directly from Top Management, if difficult decisions are to be taken. This makes solving problems easier in negotiations with the client.

The authors strongly advise to have an organisational structure as per Fig. 8.6 also in case of legal disputes, where the Monitor of Litigation (see Sect. 6.2) should be reporting directly to Top Management. If the Project Manager or Director has been reporting to Top Management during the execution of the project then Top Management will be much better informed about the claims in a legal dispute.

The direct reporting of Project Management to Top Management makes this position more attractive and thus will interest better qualified persons to apply for Project Management positions in a company. Schedule 8.6 gives a list of attributes a good

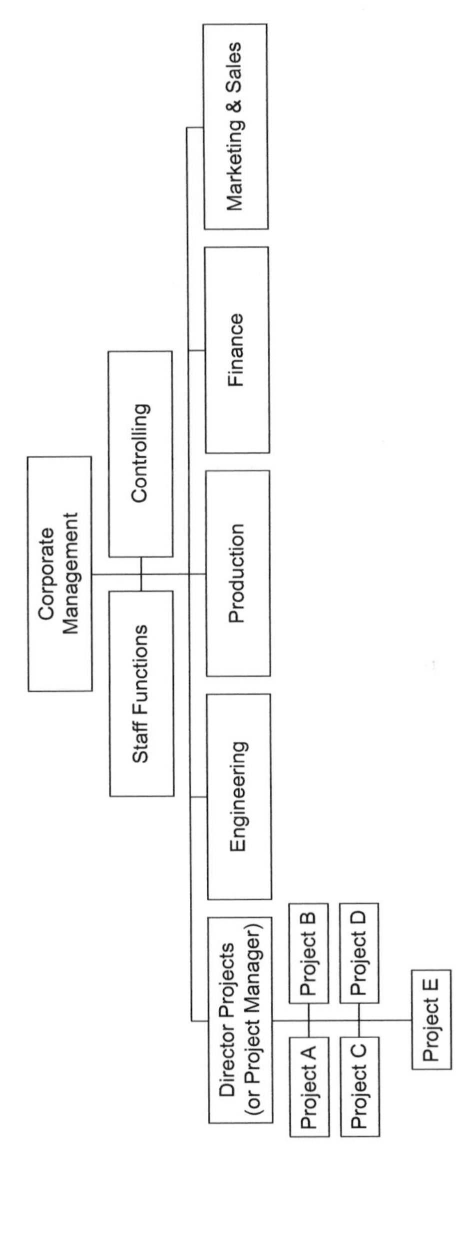

Fig. 8.6. Project Management Reports to Top Management

Project Manager should have. Since there will not normally be any person available that fulfils every attribute to 100%, we give a rating of the attributes in their order of importance with high, middle and low.

When having to chose a project manager, one can use Schedule 8.6 by giving a candidate a score for each attribute between 0 (attribute not existing) and 5 (attribute very highly fulfilled) and multiply each score by the rating in Schedule 8.6 with 3 for high, 2 for middle and 1 for low. The maximum score attainable on the basis of Schedule 8.6 is 155 points. Any person who is below a score of 100 should not be chosen as project manager.

With respect to legal disputes a Monitor of Litigation should be installed and he should not be the same person as the Project Manager of the project in legal dispute. This was discussed in detail in Sect. 6.2 above. But the attributes of a good Monitor of Litigation are practically the same as those of a good Project Manager. In fact very often the monitors of litigation have been project managers in other projects.

With respect to legal proceedings, attribute nr. 5 in Schedule 8.6 is of very high importance. In many proceedings, where the authors have been involved, it could be seen that a good record of all events during execution of the project was essential for winning their case. One can say "well documented, bingo!", meaning that the party which has all documents available for proving their case have a considerable advantage over the opponent who is less well documented. Thus, with respect to possible legal proceedings a thorough documentation is an absolute must. The good documentation also proves extremely helpful in negotiations/discussions with the client during execution of the project, when differences of views occur. It is then helpful that one can prove one's case.

Good documentation of the events during a project also improve the esteem of the client towards the contractor and it will in case of disputes make the client think twice, whether he wants to go for legal proceedings. The good documentation is something everybody says that it is necessary, but there are so many projects where

Schedule 8.6. Attributes of a good Project Manager

	Attribute	**Rating**
1.	Negotiation capacity	high
2.	Good technician in the field of the project	high
3.	Generalist with a wide technical and financial view	middle
4.	Is able to work in a team	high
5.	Well structured and organised in his work	high
6.	Is creative	low
7.	Good communicator	middle
8.	Ability to motivate people	high
9.	Ambition	low
10.	Decision maker	middle
11.	Foreign language capability	middle
12.	Managerial experience	high
13.	Ability to evaluate and select people	high

documentation is extremely poor and when it is done it is often very late. Documentation of a project is not a very popular task on project site.

8.2.2 Team Building and Team Relations

There is an old saying, which goes "bad boss, bad team, bad results", and there is a lot of truth to this statement. It is the boss who forms the team and not the other way around. A good team on the contractor side results normally in a good relation with the client's team. Or if we put it the other way around, a bad team will have problems with the client's team. Therefore it is in the interest of every team member to think "team" and to avoid "solo numbers".

A good project manager is normally also a good team builder. Nevertheless here in Schedule 8.7 we give some indications, how the project manager can go about to build a good team.

In this book we deal with conflicts with clients or subcontractors, so called outside conflicts, and not with conflicts inside the company or the project team. The inside conflicts need a different approach and use different tools than those, which are used in outside conflicts. In an inside conflict there are no legal proceedings, but there are bosses who decide, who have their special opinion and who are generally not subject to a legal code. The approach is therefore different. There are, though, some similarities, since there are humans acting as well inside the company as in the client's or the subcontractor's team, but their objectives are different. Therefore one has to keep always in mind that they are humans but with different interests.

The team building is therefore extremely important to run a project successfully. If the team inside the project functions well, then there is a good chance that the project will be a success. Many projects have been unsuccessful, though, because the

Schedule 8.7. Rules to Follow for Building a Team

1. Choose your team members as carefully as circumstances allow
2. Obtain information on the team members' previous superior
3. Define clear responsibilities and interfaces to other team members
4. Listen before giving orders
5. Give credit to team members when they made a valuable contribution
6. Apply management by walking around
7. Meet with your direct collaborators once per week
8. Inform the team on company matters, make the team feel "informed"
9. Introduce key people of your team to your client
10. Manage your team transparently
11. Make your team members feel supported by you
12. Evaluate performance of your personnel once per year
13. Let your collaborators present their results, don't put yourself too much forward
14. Install information procedures within the team upwards and downwards
15. Cut gossip and negative messages fast
16. Try to eliminate troublemakers as fast as possible if you can't turn them around

team was not in harmony. This often occurs in joint ventures, where two, three or more partners have to form the team, and where the danger is always latent, that one or the other joint venture partner follows his own interests thus not contributing to the team. Prominent examples in the recent past are the German Toll System for Trucks and the tremendous problems in delay of the Airbus A 380, where each participating country was pointing at the others as being responsible for the delay. The delay of the A 380 aircraft will and is costing Airbus many billions of Euros. The delays are in this case not due to technical problems but they are due to bad coordination and bad cooperation between the partners.

Team building is therefore a must in any project, especially in those projects which have an international team and a world wide community of clients.

8.3 Case Studies and Questions

8.3.1 The Spring Roll Machine

A) Description of the Conflict

- A contract was concluded between a Swiss machine manufacturer and an Austrian dealer of foodstuff for a set up to manufacture spring rolls out of a vegetable mix and dough. The set up consisted of two machines. In the sales contract the machine manufacturer gave a detailed description of each machine with all relevant parameters. The buyer of the machines gave the machine manufacturer some 20 kg of vegetable mix and asked, whether he could produce spring rolls with his machines. The seller confirmed that his machines could produce the spring rolls.
- The buyer gave the order for the two machines stating "Machines to produce spring rolls from vegetable mix and dough". The 40% down payment for the machines was made and after three months the machines were delivered. When the machines were installed and a test run was made, the resulting spring rolls were not usable, the vegetable mix inside the rolls were of unacceptable quality, the vegetable slices appeared ground and had lost practically all their juice.
- The manufacturer made some changes to the machines and a second test run was made. The result was the same as in the first test run. The buyer refused to accept the machines and requested that the manufacturer take the machines back.

B) The Parties' Approach to Resolution of the Conflict

- The seller of the machines requested the rest of the outstanding payment for the machines, left them at the buyer's factory but took out some parts of the machines, so that they could not be used.
- The buyer requested that the contract be declared null, the machines taken away from his premises and the down-payment be reimbursed.
- The parties negotiated several times, the buyer offered that the machines be taken away and the down-payment to be left with the seller. The seller did not accept.

- The seller asked for legal proceedings. The court called in an expert, who confirmed that the spring rolls were not usable.
- The court asked for an other negotiation between the parties, but they did not reach an agreement. The court decided that the buyer had to pay half of the contracted sum for the machines and that he could keep the machines.

C) *Comments on Lessons to Be Learned from the Case*

- The buyer and the seller failed to include a detailed task statement (performance specification) in the contract. The machines were well defined but not the objective, which had to be reached by the set up of the two machines together. It was the buyer's fault that he did not make a clear task statement, which would have specified the quality of the resulting spring rolls. He bought two machines without clarifying his objectives by just stating "machines to produce spring rolls from vegetable mix". Neither the quality of the input (the vegetable mix and the dough) nor the quality of the output was defined.
- The court could not award 100% to either party, since they both were responsible for the badly defined sales contract, therefore the court split between the parties.
- Each party lost money, time and did not get anything in the end. The seller got only 50% of his sales price and the buyer had two useless machines and had to pay 50% of the purchasing price.

8.3.2 The EPS plant

A) *Description of the Conflict*

- An EPS-plant (expanded polystyrene plant) was sold by a French manufacturer to a Swedish company, which needed the EPS slabs for insulation purposes in the construction industry. The EPS-plant was supposed to produce EPS slabs of 20 and 30 mm thickness fully automatically from pre-foamed material and to pack the slabs, so they could be directly loaded on trucks and transported to the various clients of the Swedish company.
- The contract was a turn key contract with supply of the machines, erection and testing till final handover. When the plant was finished and testing started, various deficiencies of the plant were detected, such as problems with the silos for the pre-foamed material input, problems with the hot wire cutting machine, frequent blockage of the piling machine and the insufficient reliability of the conveying system through the edge milling machine.
- The buyer requested that the deficiencies be made good. The supplier made several attempts but most of the deficiencies were not or not totally repaired. The turn key plant did not operate the way the Swedish buyer had expected.

B) *The Parties' Approach to Resolution of the Cconflict*

- The parties negotiated but did not find a solution, which was acceptable to both parties. The buyer requested 1.2 million Euros for repair of the deficiencies, which he wanted to have done by a competitor of the seller, for loss of production and

for the cost the buyer had during the long test runs and unsuccessful production runs. The seller did not accept.
- The buyer went to court and requested the above stated sum.
- The court requested an expert to analyse the various defaults of the plant and make an estimate of the cost to repair the plant. The expert estimated the repair cost in total to 202 000,– €. These costs were actually estimated lower, but the seller was in various instances unable to provide the necessary proof that some of the deficiencies were due to insufficient maintenance by the buyer. Certain repairs had to be charged to the responsibility of the seller, because of his lack of documentation.
- The court decided, on the basis of the expert's report, that the buyer could reduce the purchasing price of the plant by 250 000,– €, taking into account the cost for the repairs and the cost the buyer had incurred during the testing and insufficient production time. The buyer was free to repair the defaults of the plant on his account.

C) Comments on Lessons to Be Learned from the Case

- The supplier had actually delivered a technically good plant, which had some minor adjustment problems in the various sections of the plant. His insufficient documentation, reporting and recording of the events during erection and testing of the plant put him into a defensive role, which finally made him loose a substantial amount of money.
- The buyer went immediately to court instead of trying to find an amicable solution by using an outside expert. It was finally the expert who decided the case, which could have been the case without going to court.
- The cost for the lawyer, the court and the in-house cost of the buyer where considerable and amounted to approx. 130 000,- €. An amicable solution would have been more advantageous for the buyer than going to court.

8.4 Conclusions to Chapter 8

It has been shown that the various existing tools of Project Management will avoid or reduce the risk of conflicts in international projects, when they are correctly used and consistently applied. The establishment of a Project Plan before deciding to go ahead with a project is the appropriate basis for Top Management to decide, whether to go ahead with a project or not.

The project plan with the detailed work break down structure, with the time analysis on PERT basis, with a budget planning related to the work breakdown structure, allowing a serious cost control throughout the project, and with a responsible risk analysis are the basis for successful project execution. Using the stated project management tools will also provide a solid basis for the cooperation between client and contractor, it will improve the contract drafting (Chap. 3) and avoid conflicts and legal proceedings between them.

Chapter 8 with the case studies presented has shown that going for legal proceedings, when conflicts arise, is not always a must and, that the tools and suggestions made, help to find amicable solutions, which maintain a good relationship between the parties and thus offer future business opportunities between them.

8.5 Questions on Chapter 8

1. ***Work Breakdown Structure***
   Add the following work packages to the work breakdown structure in Fig. 8.1 and give them numbers: *Excavation, Concrete Works, covered concrete tanks, Piping, Pipe Racks, Sea Water Pump Motors, Recirculating Brine Pump Motors.*
2. ***PERT-Analysis***
   Can the total slack of an activity in a PERT-schedule of a project be totally used to recover delays, which occurred in an other activity, or have other measures to be searched for in order to recover the delay?
3. ***Critical Path***
   Determine the critical path in the project of Fig. 8.3, calculate the length of the project and compare to the time scale at the bottom of Fig. 8.3.
4. ***Virtual Delay***
   Try to explain the term “virtual delay”.
5. ***Responsibility Matrix***
   Establish a task responsibility matrix for the project “Erection of a prefabricated family house”, as shown in Fig. 8.3.
6. ***Gantt Schedule***
   Which advantages has a PERT schedule over a GANTT schedule and vice versa?
7. ***CE-Standard***
   How can a machine, constructed according to Japanese standards, be imported into the European Community where the CE-conformity is a must?
8. ***Risk Analysis***
   List the potential risks of a project for a bridge crossing the Gibraltar strait between Europe and Africa, group the risks and estimate the risk probability.
9. ***Project Manager***
   Evaluate a project manager of your choice with respect to the attributes in Schedule 8.6, using the quantification rule given in Sect. 8.2.1. In a second step evaluate your own qualities as a project manager and compare the overall score with that of the manager chosen.
10. ***Team Building***
    Classify the 16 rules given in Schedule 8.7 by rating them of “high”, “middle” or “low” importance.

References

[1] Eisner H (2002) Essentials of Project and Systems Engineering Management, New York, USA.
[2] Daenzer W F (1977) Systems Engineering, Köln, Germany.
[3] Felding F (2005) Senior Project Manager Course, Maersk Training Centre A/S, Svendborg, Denmark.
[4] Henley E J, Kummoto H (1981) Reliability Engineering and Risk Assessment, New Jersey, USA.
[5] Turner J R (1993) The handbook of Project-based Management, McGraw Hill.
[6] Elkjaer M, Felding F (1999) Applied Project Risk Management, International Project Management Journal Vol. 5, No. 1, Finland.

9 Conclusions and Recommendations

In this book we have tried to answer the question: "What can be done to prevent, resolve or reduce conflicts in major international projects?" in order to minimize the inefficiencies, the trouble, the loss of money and other mischief they cause.

The main conclusions and recommendations to achieve the reduction of conflicts are the following:

- The Contracts of the projects need to be improved with regards to the following aspects:
  1. Clarification of the obligations due to the degree of turn-key responsibility and the "fit for the purpose" principle
  2. Definition of Owner's supplies, services, battery limits and interfaces with the contractor's responsibilities
  3. Split of responsibility for regulatory aspects of plant design, construction and operation
  4. Improved technical specifications leading to fewer interpretation disagreements
  5. Contract provisions for pre-litigation mediation between the parties by a third party mediator
- Conflict prevention and handling can be improved by proper application of Psychology as follows:
  1. Behind every project conflict lie psychological issues, they have to be analyzed and handled
  2. Project Management needs to apply the psychological tools presented in Chap. 4
  3. Understanding of conflict development stages, dimensions, and different personality types are essential
  4. The benefits of employing a psychological approach in project management are better working relationships and reduction of conflicts
- Negotiations to resolve conflicts can be improved by:
  1. More emphasis on building good relations between the contract parties
  2. Early detection and professional handling of disagreements before developing into a conflict
  3. Better and deeper understanding of the background leading to a conflict
  4. Careful preparation of conflict resolution / settlement negotiations
  5. Improved creativity and professionalism in negotiations to reach a settlement

- Litigation can be made more efficient by the following measures:
  1. Nomination of a Monitor of Litigation, who is not the Project Manager
  2. Development of a sound strategy to be followed during litigation
  3. Resolving conflicts through mediation or other "soft" resolution methods instead of court trials
  4. Analysing the claims thoroughly, remembering that "a bad compromise is better than a good judgement"
  5. Avoiding the high cost of a trial through careful cost analysis of legal proceedings before going to trial
- Handling of technical expertise (prior to litigation or as part of litigation) can be improved by observing the following recommendations:
  1. Choosing the right expert as from the beginning
  2. Giving the expert the opportunity to mediate between the parties
  3. Analysing and choosing the right strategy to obtain a favourable expertise
  4. Trying to reject an unfavourable expertise through rejection of the expert
- Conflicts can be avoided or resolved by employing the following essential project management tools:
  1. Establishing a detailed project plan before implementation of the project, including a detailed risk analysis and contingency plan
  2. Defining and using a detailed work breakdown structure during project execution
  3. Controlling the project execution by employing a PERT-type time management system, based on the work breakdown structure with a bi-weekly update
  4. Installing a clear organisational structure, agreed upon by client and contractor, with a well defined responsibility matrix
  5. Controlling the project with a budget control system based on the work breakdown structure leading to regular invoicing and updating with the client
  6. Documenting project progress on the basis of a reporting system, which has been agreed between contractor and client
  7. Applying a final acceptance procedure, which has been agreed upon between client and contractor

The issue is in our opinion more complex than just implementing each and every of these recommendations. International projects with fixed prices and firm delivery dates need a sound system of basic contract templates beyond the ones, which already exist.

New principles should be introduced, in order to achieve a better process design and related engineering work on one side and a contracting structure with a clearer division of responsibility and an early contract drafting on the other side, in order to break the unfortunate increase of conflicts in international projects.

In our opinion the problem is that there seems to be a gross underestimation of the resources required for the many agreements during project implementation. Some project people believe that practically the only real agreement is associated with the contract negotiation and signing procedure. This is very far from reality. The contract negotiation might be the first agreement, but many have to follow, ending with the final agreements to transfer the plant to the Client and settling the final accounts.

Signing of the contract means that the parties have reached an agreement regarding the basic framework and the specifications for the execution of the project. At the same time they have frozen the price and time of delivery. The conflicts then start during execution of the project, because some Contractors think they can get away with substandard work and some Clients believe they can improve specifications and scope of work beyond what was agreed upon, without compensating for the extra work. Watch out for those who misuse the system!

Primarily during the engineering phase but also during installation and commissioning a significant degree of flexibility and "give and take attitude" is required by both parties to reach an agreement on the adjusted scope of work. It is in the interest of both parties to reach a settlement because it saves time and money for both. What we often experience, though, in major international projects are long struggles led by "single minded bureaucrats" because management believes that the project teams should just stick to their own specifications, time schedules and budgets. This attitude will often result in conflicts and delays and leasding eventually to litigation, which costs a fortune to both parties.

Dr.-Ing. Wolfgang Spiess
Consultant and sworn expert

wolf.spiess@t-online.de

Finn Felding,
Project Adviser

ff@finnfelding.dk

Glossary

Glossary of principle terms used in the book with a short explanation, the terms are included in the Index, where the pages are given, to find the terms treated in the text of the book.

Acceptance. See Plant acceptance

Activity plan (and project time schedule PERT). A list of activities in the order of interdependence, where each activity has a starting event and a completing event. The whole activity plan and time schedule is presented as a chart showing the interdependence or as a list of activities)

Advance payment bond or prepayment bond or down payment bond (bond can be substituted by guarantee). Financial guarantee to cover for Contractor's inability or unwillingness to refund the down payment in case he/she stops performing the Contract

ADR. Alternate dispute resolution, soft method to resolve a conflict instead of or before undertaking legal proceedings

Allotment. Apportionment of funds or similar

Amendment (to the Contract). An agreed addition to the contract, formally made as a supplement according to the same procedure as for agreeing and signing (and make effective) of the original Contract

Arbitration. Private (no disclosure) litigation organized by international organizations as for instance the International Chamber of Commerce and agreed between the parties in the contract as the contractual procedure and method for dispute resolution should negotiation fail

Arbitrator, Arbitral Tribunal. Independent, unbiased and qualified persons appointed by the parties and by the arbitration institutions to form the Arbitral Tribunal

Arbitration clause. Clause in the contract of a project, which defines when and how an arbitration tribunal may be called upon by the parties in case of conflict

As built drawing (or documentation). Drawing corrected for changes during installation and commissioning, showing the plant as it was finally built

Bar-Chart. Plan which assembles the various activities of a project above a time-scale, thereby showing the length of the activities and the whole project. Also referred to as "Gantt-Chart"

Basic (process) design. The engineering work to establish the process configuration, the main design parameters, the general lay-out and the performance criteria

Battery limit. Limits of the area occupied by the project, sometimes used as limit between the areas of responsibility

Biased. The opposite of neutral and objective

Bid. The proposal and quotation submitted by the supplier or contractor invited to quote according to the tender procedure and requirements

Bid bond or Tender Guarantee. Financial guarantee to cover for Contractor's unwillingness to sign the Contract if awarded

Budget control. Control of spending versus budget on item level, calculating and explaining deviations. The budget control also includes general recommendations on how to regulate the contingency amount, set aside in the budget, and forecasts the expected financial outcome of the project at completion

CFR. Delivery "Cost and Freight at NN" (place where the vessel is unloaded with the goods or equipment – name to be inserted) – INCOTERMS 2000 condition of delivery

Civil contractor. A contractor or subcontractor responsible for earth mowing, site establishment, foundation work, internal roads, building work etc.

Claim. A legal or contractual written request for compensation of an event to the party causing the event

Commissioning. Bringing equipment into operation. Used for all activities after start-up of the plant with raw materials to fine tune the operation, prove the performance and fulfill the contract until Acceptance

Conflict. A fight or struggle

Conflict contingency plan. Policy, procedures and rules regarding spotting of disagreement evolving to a conflict and how to solve the conflict as early as possible

Consultant. A third party hired by one of the parties to render professional services (e.g. in engineering, construction or project management etc.) to the principal

Contract execution. Fulfilling all the obligations of the contract

Contract interpretation. A technical and practical description of what is needed according to the technical requirements as found in the contract, often in various sections and articles

Contract Language. Language in which all communications of a project have to be expressed, the language has to be stipulated in the contract

Contract Procedural Law. The applicable law which governs the legal procedures, it is not always identical with the contract law

Contract Law. The applicable law which governs the contract, should not be confounded with procedural law, which may but does not have to be different from the contract law

Contract provision. An article in the Contract. The word is used when reference is made to contract articles or provisions in general

Contract specification. The technical description incl. specific technical requirements as found in the contract, often in various sections and articles

Contractor or main contractor. The company carrying out the work required in the contract

Contractor's All Risk insurance (CAR insurance). The name of the insurance policy which covers nearly all insurable risks the contractor may suspect during execution of the contract, incl. the risk of the plant before Acceptance. The CAR insurance does not cover the professional design liability (professional indemnity – separate insurance, normally not attractive compared with the risk occurrence)

Contractual. In accordance with the contract incl. annexes etc.

Corporate Pledge Model. A basic statement in a company's General Conditions, that the company would, in case of conflicts, first try to resolve the conflict by mediation instead of going to court immediately

Cost-benefit analysis. A comparative analysis of different solutions or methods where the difference in costs are compared to the difference in benefits (measurable and non-measurable) in order to establish which alternative gives most "value for money"

Cost of Expertise. The cost to be paid for an expert

Cost of Arbitration. The cost to be paid for the arbitrators and for the administration of the body, which governs the arbitration

Court or State Court. The ordinary courts of a given country

Critical Path. The conjunction of activities of a project, which determine the longest duration of execution, therefore the end of the project is at the end of the critical path. The critical path has no slack, and any delay on the critical path provokes a delay of the whole project

Customs clearance. The actual approval by the Authorities of the project country that the equipment etc. are released by customs for transport to the construction site

Delay analysis. An analysis of the causes for a delay and of the contractual justifications for each delay, with the objective to reach a conclusion on how the total delay should be distributed between the parties

Delay causing event. A physical event (or lack of event) causing one contract party to be delayed, compared to the official project time schedule

Delay claim. A claim for extra costs caused by the other party in being late in her performance

Delay. Actual delivery takes place at a later time than agreed in the contract or a confirmed order. The difference in time between the actual time and the agreed time is the total delay for which responsibility has to be distributed between the parties

Delay Penalties. A fine (mostly in money) levied by the Client on the Contractor for late execution of a task

Disburse. Providing cash money

Dispute Review Board (DRB). US Third Party Inspectors

Early Warning System. Reporting System to Top Management on details of project execution in order to detect possible conflicts at an early stage with the intention to take early mitigation measures

Effective contract. A contract where the parties have fulfilled all the agreed provisions regarding bank guarantees, down payment, import license, etc. in order to make the contract operative

Engineering approval. The Owner's (or Employer's or the Engineer's) acceptance that the engineering complies with the contract specifications

Engineering, Supply and Construction projects/contracts. See turn-key

Environmental impact assessment. Analysis and evaluation of all the environmental consequences of a new plant (or extension of an existing plant)

Evidence. Indisputable records of what actually happened relevant to the case of disagreement or conflict

Expert (in relation to project conflict). An unbiased third party invited by the parties or the Court to give his opinion on certain issues described in his Terms of Reference

Extra cost claimed. The alleged extra costs which one party believes the other party should carry in accordance with the contract provisions

Factory Acceptance Test (FAT). The buyer's inspection and testing of the equipment and its documentation ordered to establish its conformity with the agreed specifications

FCA. Delivery "Free on Vessel at NN" (place where the vessel is loaded with the goods or equipment – name to be inserted) – INCOTERMS 2000 condition of delivery

Financial guarantees (other than payment guarantee). Guarantees issued by banks or insurance companies to cover Owners' losses due to bankruptcy, insolvency or inability of the contractor to perform his work

Financial Institution. A public institution or a private limited company that borrows money on the market and lends it out (e.g. to finance projects)

Financial obligations. The obligations in a contract regarding financial guarantees, payments, taxes and duties, reimbursement of certain costs and other financial matters

Financing agreement. The agreement between banks, financial institutions and Owners on how to finance a project. It is completely separate from the Contracts to supply the equipment and perform the construction and installation work

Fit for the purpose. FIDIC: (Conditions of Contract for Design-Build and Turnkey 1996 p. 41, first paragraph (4.1)) specifies that the work shall be fit for the purpose, and that the purpose is ascertainable from the Contract

Fixed price (contract). A contract stipulating a final, firm, lump-sum, non-revisable price except regarding extra work and claims agreed between the parties

Functional guarantees or performance guarantees. See performance guarantees

General and Special Contract Conditions. Client's standard contract conditions. General Conditions apply to all projects of a specific client, whereas Special Conditions are specific for each project

ICC Arbitration. The arbitration institution of the International Chamber of Commerce in Paris, France

ICC Court of Arbitration. The ICC body controlling the arbitration cases and tribunals under the ICC Rules of Arbitration

Import license. The approval by the project country Authorities that certain equipment etc. can be imported under certain conditions

Incompetence. Lack of competence to carry out a specific work. Competence means sufficient skills and experiences required by good practice in the trade

Independent proof procedure. to establish missing evidence for a court case or for arbitration

Inspection Authority. Third party organization hired by the parties to perform inspection of equipment at origin, before shipping and on site of a project to ensure that the equipment and the work comply with the contract specifications

Installment. Part payment – See payment conditions

Instruction (according to Contract). The Owner's order to the Supplier or Contractor to carry out a certain work etc. in accordance with the Contract. The Supplier or Contractor has to adhere and comply right away – objections are handled as a dispute

Justification (contractual). The legal reason, according to the law and the contract, why a certain deficiency in the work or supply is the responsibility of the other party

Lay-out drawing. An overall drawing of a project, showing the positioning of equipment, material, foundations and the connections between them

Legal adviser. Lawyer for one of the parties advising his/her principal

Limits of supply or limits of work or battery limits. points or lines, where the responsibility changes between one party to the other (owner to contractor, contractor to subcontractor)

Liquidated damages. The compensation fpr non-fulfilment of a delay or function

Litigation strategy. Plan of action or policy for a litigation

Litigation. Proceedings at a State Court

M&E contractor. A contractor or subcontractor responsible for the mechanical and electrical installation of process equipment and of supply lines, auxiliary utility systems as fire protection and fire fighting, Heat, Ventilation and Air Conditioning (HVAC) etc.

Maintenance manual. A manual describing in detail how to maintain the plant, how to detect malfunctions and how to handle safety issues

Managing Director (MD). Head of a company or Division thereof

Material obligations. The obligations in a contract regarding engineering work, governmental approvals, supply of equipment, shipping, construction, installation and commissioning work

Minutes of Meeting (MoM). A written and signed off recording of the agenda of a meeting, its participants and decisions

Mediation. Soft resolution negotiations facilitated by a private expert as mediator

Mediation Clause. Clause in the contract of a project, which defines when and how a mediation procedure may be initiated by the parties in case of conflict

Monitor of litigation. An experienced person appointed by a company, parallel to the Project Manager, with special responsibility for analyzing the conflict, searching for solutions, conducting negotiations and managing litigation

National Chamber of Commerce. The national organization representing the business community in a country

Negotiations. Direct discussions between the parties involved in a disagreement with the aim of reaching a commercial compromise satisfying both parties and characterized by the "give and take" attitude

Network analysis. Analysis with respect to time and cost of a project by analyzing the project elements and their interdependencies

Notification (of claim situation – extra cost and/or extension of time). A written notice from the party suffering the delay to the party causing the delay stating the facts of the delay and its potential consequences in broad terms

Ombudsman. An official appointed by a government to investigate individuals' complaints against public authorities (a word of Scandinavian origin)

Operation manual. A manual describing in detail how to operate a plant, how to deal with malfunctions and how to handle safety issues

Operative contract or effective contract. Contract, which has been made operative after the parties have met all their obligations for the start of the project

Owner. The owner of the plant or infrastructure, which is being built or extended through the project

Pacta sunt servanda. Latin – "Contracts must (always) be complied with (in all respects)"

Parties (to the contract). The parties, client and contractor, which have agreed to work together and have consequently signed a contract for execution of the contract and payment of the related works

Payment conditions. When and how payment installments fall due. The division of the total contract price into a number of installments that fall due at different stages and milestones of the plant implementation

Payment terms. How, when and by which method and documentation a payment has to be effectuated once it is fallen due

Penalty. Surcharge which is levied mostly for non respect of delays or performance in a project

Performance bond or completion guarantee. Financial guarantee to cover for Contractor's inability or unwillingness to repay the remaining funds in case he/she stops to execute the Contract

Performance guarantee. The contractor's guarantee that the equipment or plant will reach the characteristics foreseen in the contract

Permit (or approval or license) issued by the Authorities. Authorization to go ahead with the project according to the terms and conditions of the Authorization

PERT diagram. PERT is an abbreviation which stands for Program Evaluation and Review Technique. This technique allows to analyze a project with respect to time of completion by scrutinizing the various project activities and their interrelationship (see also "Activity Plan")

Plant acceptance. The Owner's written agreement that the plant has been engineered, supplied, constructed, installed and commissioned in accordance with the contract. This includes the use of the plant by the Owner and the acceptance of the risk of the plant (or parts thereof) by him

Pre arbitral dispute resolution. Attempt to settle a disagreement or conflict prior to litigation at court or by arbitration

Pre-commissioning. Used for all activities prior to start-up of the plant with raw materials to ensure safe operation at the plant start-up

President of the Arbitral (or Arbitration) Tribunal. The Chairman of the 3 arbitrators

Prevention of conflict. Efforts by one or more parties to avoid that disagreements evolve into a conflict by solving the disagreements

Price specification. Detailing the total contract price into plant sections and items

Procurement. Activity of purchasing the equipment, materials, services and supplies for a project

Professionalism. By and large equal to professional competence: See under incompetence

Progress measurement. To establish the progress (how much work has been performed) in % of the total anticipated work for the implementation of the project in accordance with an agreed system/method on how to measure the progress

Project conflict. The result of serious and protracted disagreement(s) threatening the cooperation between the project parties

Project description. A short description of the planned project incl. owner, location, functions and processes, time schedule and budget made in order to present the project to authorities, financial institutions, investors, suppliers and contractors. A comprehensive model can be found in www.ifc.org

Project Director or Director of Projects. The superior of Project Managers

Project implementation. The activities necessary to carry out the project according to the project plan and fulfilling the project objectives

Project Manager. The manager responsible for the project implementation in all aspects

Project Objectives. The compilation of the objectives the Owner of a project wants to achieve with the project

Project Plan. A comprehensive plan in the form of a manual on how to prepare, bid for and implement the project

Project Reporting. The plan how the contractor has to report on the development of the project to the Owner. The Project Reports are structured in line with the relevant provisions of the contract

Project schedule (or time schedule) and Activity Plan. See "Activity plan"

Project. A combined set of activities aiming at the same objective with a scope of work and a complexity that necessitates a comprehensive coordination and control. The activities will typically comprise engineering, equipment manufacturing and delivery, construction, installation, commissioning and related documentation and will often take place at different locations

Proposal Manager or Bid Manager. The manager at the contractor's responsible for the project plan, bidding and contract drafting

Psychological approach (to conflict solving). The analysis and understanding of the personalities and their behavior involved in the conflict and its resolution

QA stands for Quality Assurance. Procedures and activities which establish, how the quality of the equipment, materials, works and services are to be obtained

QHSE stands for Quality, Health, Safety and Environment. Procedures and activities which plan for and establish whether the project activities are in accordance with local laws, regulations and internal standards with regard to health, safety and environmental issues

Raw material availability and quality survey. An investigation of the quality and quantity of raw materials available for the project

Referee. As in a sports game, the referee indicates in a project who is right and who is wrong, the parties are free to accept the verdict or reject it, unless otherwise foreseen in their contract

Regulatory work. The work in a project resulting from laws, regulations, standards

Reimbursement. Covering the cost as a refund of the other party as per agreement

Responsibility Matrix. Matrix, which assigns tasks (responsibilities) to persons or entities, can be applied within a company or a project, defining in the latter case the responsibilities of the Owner and of the Contractor

Retention bond or Warranty guarantee. Financial guarantee to cover for Contractor's inability or unwillingness to refund the extra cost to change supplier/contractor in case he stops executing the Contract or comes into default

Risk Analysis. Activity to identify possible risks of a project

Scope of work or scope of supply. All what has to be "delivered" (equipment, materials, spare parts, documentation, works and services etc.) according to the contract

Senior Management. Management in a company as from a certain level of authority upwards (normally as from department heads upwards)

Settlement agreement. A written signed and executable agreement between the parties settling a contested issue

Settlement proposal. A proposal issued by one of the parties or by a mediator/expert of how to settle a disagreement or conflict

Settlement. A written agreement signed by all concerned parties, containing clear and executable provisions for settling the disagreement

Single line diagram (electrical-). A diagram showing the basic electrical distribution system from the public grid to the control centers incl. main transformers and switchgears

Soft resolution method. Mediation before litigation

Soft resolution. Dispute resolution by negotiations between the parties, facilitated by a private expert or mediator

Soil investigation. Drilling, sampling and analyzing of the underground in order to assess the soil's suitability for supporting structures necessary for building the plant

Solution of conflict. Efforts by one or more parties to reach a settlement agreement which solves a conflict

Specifications. The general term for the task statement of a project

Specification of raw material. Technical description of the raw material which a process requires in terms of quantity and quality and especially which raw material the process can not tolerate

Specification of utilities. Technical description of the utilities which the process requires in terms of quantity and quality

Split of obligations. Division of responsibilities for activities necessary for designing and building of the plant

Subcontractor, sub-supplier. A company providing engineering, equipment/material or services to the main supplier or main contractor

Strategy of Litigation. The compilation of actions planned to be taken in the case of legal proceedings (litigation or arbitration)

Supervision (of installation work and commissioning). Carried out by the supplier's supervisors who advise the Buyer (or Owner) about the correct installation, erection, assembly, start-up and operation of the equipment and the whole plant as agreed in the Contract or in a separate Supervision Agreement

Supplier, main supplier or process supplier. The company designing, manufacturing and delivering equipment and process line incl. documentation

Task Responsibility Matrix. see Responsibility Matrix

Task Statement. The statement of the Owner of a project, where he states the objectives to be achieved and the tasks to be executed in the project

Technical annexes (to the contract). Attachments to the legal part of the contract ending with the signature. Mostly containing technical specifications and procedures. Technical annexes (as well as other annexes) form an integral part of the contract and are signed by the parties

Tender documents. The Owner's task requirements and conditions, upon which the bidder may make his offer

Terms of Reference. Formal description of the task for an Arbitration Tribunal or an Expert as basis for their work

Test run. Agreed and controlled plant or equipment operation for the purpose of proving certain guaranteed performance parameters

The Engineer. The technically responsible person appointed by the Owner of a project to control technically all matters of a project. Normally the contract specifies the Engineer's duties and authority (approvals, instructions, variation orders, extension of time etc). Basically the Engineer's decisions must be followed in the day-to-day project work

Third Party Inspector. Inspector installed and normally paid by the owner of a project to control the quality and the quantity of equipment and goods to be delivered to the project

Time extension claimed. The alleged extra delivery time which the supplier/contractor believes the Owner should grant, when extra work is requested by the owner or other circumstances lead to prolongation of execution of the project

Time schedule. See "Activity plan"

Turn-key contract. A design and build contract with fixed price and firm stipulated delivery time (plant acceptance)

Variation order (or Change order or Extra Work Order). An order by the Owner (or Employer) to carry out extra work (outside the work stipulated in the contract) at an agreed extra price and an agreed extra time

VAT (Value Added Tax). The Tax which is levied in a country on top of the sales price, which can be partly or totally reimbursed in export dealings

Virtual Delay. Delay, which occurs, when parallel activities are in delay, for which different parties are responsible. The virtual delay is that part of a delay for which both parties are responsible and none of them carries the whole responsibility

Warranty claim. The Owner's claim to the Supplier or Contractor to repair a defect or improve the plant performance

Warranty or Warranty period. The supplier's or contractor's warranty (guarantee) against defects in material, workmanship etc. during the warranty period of normally 24 months

Work break-down structure (WBS). A systematic partition of the project in reasonable and workable supply and work packages for the purpose of scheduling (PERT), task repartition among the parties, budget control and invoicing

Index

Acceptance procedure 144
Administrative fee 115
ADR
  techniques 109
ADR-Clause 58
Agreed Time Frame 17
Agreement 38, 80, 81
America 28, 36
American Arbitration Association 110, 127
Amicable settlement 130
Analysis of Position 80
Appointment of an expert 126
Arbitration 72, 105, 110
Arbitration commission 112
Arbitrator 112
  - the choice of 113
Asia 28, 34
Attitude 79
Authorities 11, 12

Bar-chart 147
Behaviour 79
  - destructive 55
Bid 38
Break-down of Negotiation 82
Budget 154
  -plan 154
Budget control 100

CAR 19
Case Lesson 90, 92–95
Case Study 88
CE - standard 167
Centre of gravity 47
Change 63
Change in Quantities 67
Change orders 144
Checklist
  - psychological 59
Chinese law 107
CIETAC 111
Claim 70, 73, 80
Claim Evaluation 75
Client 19
Commissioning 16, 18, 157
Communication 78
  - constructive 55, 56
  - destructive 55, 56
Completeness 17
Compromise 109
Conflict 38
  - competent leader 2, 54
  - constructive responses to 54
Conflict Cause 64
Conflict competency 43
Conflict Contingency Plan 9
Conflict Development 65, 73
Conflict escalation 45, 46
Conflict Resolution 77
Conflict Source 64, 66
Construction 17
Construction Site 72
Contract 16, 23, 25, 38
  - psychological 58
  back to back - 153
  types of 1
Contract Function 28
Contract Interpretation 73
Contracting Structure 6, 7, 9
Contractor 11, 23, 64
Contractual Conflict 30–32, 71
Contractual Position 73
Controller 152
Corporate pledge model 108
Cost
  in house 104, 116

of an expertise 133
of a lawyer 116
of a state court 104
of litigation 114
Cost-benefit Analysis 70
CPR 109, 111
Critical path 150, 167

Dalai Lama 54
Decrees 10
Delay 16, 22, 23, 28, 67, 84, 85, 96
- control 152
- penalties 153
virtual - 152
Delay Analysis 86
Design 18, 64
Differences
- organizational 57
Dimensions
- instrumental 46
- of interest 46
- personal 47
-of value 47
Disagreement 64, 65, 67
Dispute Resolution 17, 27, 96
- alternative 58
Distribution 66
Documentation 162
Drafting 28, 34, 36, 81
Drafting Process 17
DRB Foundation 107
Dynamics of conflicts 43

Earliest end date 150
Early warning system 100
Engineering 16–18
Environment 9, 18
Environmental instructions 11
Equipment 16–20, 23, 26–28, 31, 40
Erection 16
Estimation of the Outcome 80
Europe 28, 36
Evidence
preparation of 118
Exclusion
from site inspection 131
Executing the Agreement 83
Expert 113, 126
- generalist 127
- hourly rate 134
- impartiality 132
- incompetence 133
- nominated by a court 127
- outside expert 104
- rejection of 132
- release of 133
- specialist 127
Expert's capabilities 127
Expert-witness 104
Expertise 118
- cost of 133
- execution of 130
Extension
of expertise 131
Extension of Time 84, 96
Extra Cost 85

Fee
for an arbitrator 115
FIDIC 37, 39
Finance Agreement 25
Financial 65, 67, 73
Financial Guarantee 25
Financial Obligation 17, 23
Financial Risk 24
First Initiative 77
First Negotiation Meeting 78
Fit for Purpose 16, 20, 41
Fixed Price 15–18, 20
Force Majeure 22
Friedrich Glasl 43, 44
Functional guarantees 143, 144

GANTT
- schedule 167
Gantt chart 146
General Conditions 38
Generalist 129
Gross negligence 133

Health 9, 11, 18

ICAC 111
ICC 37, 110, 127
- International Chamber of Commerce 106
INCOTERMS 20, 37, 39
Independence

of arbitrator 114
Independent expert 103
Interests 6, 10
Interface 18

Joint venture 164

Key Staff 72

Laboratory
in an expertise 132
Language 156
Laws 9, 154
Level
- of detail orientation 53
- of dominance 51
- of intensity 53
- of the extrovert 52
Litigation 32, 69, 72, 109, 113
simulation of 103
strategy 103

Main Contractor 6–8, 63, 69
Main Process Supplier 6–8
Management 73
Marguerre 57
Material Obligation 17
Mediation 33, 106
Mediator 107
Mistake 63
Monitor of Litigation 101, 160

Negotiating Team 75
Negotiation 32, 68, 71, 73, 77
Negotiation Meeting 79
Negotiation Parallel 83
"New York Convention" 111
NGO 12
Nomination of an expert 126

Objectives 143
ORGALIME 37, 39
Organisational structure 160
Outside expert 103
Owner 7, 11, 63

Parallel contractors 7
Part 9
Partiality 114
Parties 1
Party 5, 6, 8, 12, 64, 73
Party Position 78
Party's expert 126
Payment Security 24
Payment Term 23
Penalties 144
Performance Guarantee 17, 18
Personality
- categories of 51
- indicators 51
- types of 50
PERT 146, 167
- chart 146
Practise 37
Pre-arbitral rules 106
President
of the arbitral tribunal 114
Prevent 31
Prevention 32, 68, 72
Price 23, 67
Procedural law 112
Process Supplier 11, 63, 64
Procurement 153
Project
- acceptance procedure 156
Project country 9, 11
Project Culture 72, 87
Project Implementation 63
Project management 141
Project manager 99, 100, 160, 167
Project Organisation 17, 27
Project Plan 9, 142
- cost of 142
- elements of 142
Project team 153
Projects in conflict 1
Provisions
- client's 142
Psychological aspects 2
Psychological contract 2
Psychology
- impact of 41
Punch list 156

QHSE 9
Quality 9, 18

Reactions to conflicts 49

Referee 107, 109
Referee procedure 107
Regional Difference 34, 87
Regulation 10, 18
Regulatory Obligation 17, 22, 63
Reporting 155
  - of project management 160
Reporting schedule 156
Resolution Phase 77
Responsibility 7, 12, 16, 20, 23
Responsibility matrix 167
Retaliatory Cycle 55
Risk
  - analysis 157, 167
  financial - 159
  organisational - 158
  scheduling - 159
  technical - 159
Risk Transfer 19

Safety 9, 11, 18
Scope of Work 7, 8, 17, 18, 63, 65, 66
Secret
  of a company 130
Security System 25
Settlement 69, 77, 78, 80
Settlement of Disagreement 96
Signing the Agreement 82
Site inspection 131
Slack 150
Soft resolution methods 106
Special Condition 38
Specification 17, 65, 143
Staffing 153
Standard 10, 18, 28, 37, 39, 154
State of the art 133
Strategy
  for litigation 102
Subcontractor 69
Sum
  in dispute 115, 116
Supply Contract 16
Survey 30

Task responsibility matrix 153
Task statement 143
Tax 17, 26
Team
  - building 163, 167
  - relations 163
Tender Documents 38
Terms of reference 112
  of an expertise 130
Testing schedule 157
Third party inspector 107, 137
Time
  - control 152
Time Extension 85
Time Schedule 18, 67, 73
Toll system 159
Tools
  of project management 141
Top management 143
Trust 78
Turn-key Contract 16, 19, 22

Variation 66
VAT 134
Virtual delay 167
VTAK 111

Work breakdown structure 144, 154
Work package 154, 167